STUDY GUIDE
AND SOLUTIONS MANUAL
to Accompany

FUNDAMENTALS
OF BEHAVIORAL
STATISTICS

STUDY GUIDE
AND SOLUTIONS MANUAL
to Accompany

FUNDAMENTALS
OF BEHAVIORAL
STATISTICS

SEVENTH EDITION

Richard P. Runyon

Audrey Haber

McGraw-Hill, Inc.
New York St. Louis San Francisco Auckland Bogotá
Caracas Lisbon London Madrid Mexico Milan
Montreal New Delhi Paris San Juan Singapore
Sydney Tokyo Toronto

Study Guide and Solutions Manual to Accompany
Fundamentals of Behavioral Statistics

2 3 4 5 6 7 8 9 0 SEM SEM 9 0 9 8 7 6 5 4 3 2 1

ISBN 0-07-054328-3

This book was set in Zapf Book Light by Pat McCarney.
The editors were Jane Vaicunas, Maria E. Chiappetta, and James R. Belser;
the production supervisor was Janelle S. Travers.
The cover was designed by Wanda Siedlecka.
Semline, Inc., was printer and binder.

Contents

PREFACE

Each chapter in the *Study Guide and Solutions Manual* corresponds to a chapter in the text *Fundamentals of Behavioral Statistics*, Seventh edition. The Study Guide has been designed to provide you with review and feedback concerning your mastery of the textual materials. It is suggested that you study the assigned chapter thoroughly and complete the exercises in the text before proceeding to the Study Guide.

Each chapter in the Study Guide contains several different units. First, there is a statement of behavioral objectives, often broken down into two categories—conceptual and procedural objectives. If you achieve these behavioral objectives, you have gone a long way toward mastering the fundamentals of statistical analysis. The statement of behavioral objectives is followed by a comprehensive chapter review that covers most of the important terms, symbols, concepts, and procedures found in each chapter. A feature new to this edition is the incorporation of step-by-step computational procedures within the chapter review. Next, there follows a series of selected exercises that are designed to provide experience with the practical application of the statistical concepts and procedures introduced in the text. Detailed answers are provided to each of these exercises. Finally, there is a self-quiz composed of true-false and multiple-choice items. This unit will provide you with an opportunity to practice taking exams. It may also serve as a diagnostic test that reveals areas of weakness that require additional attention.

Special attention was paid in this edition to the Selected Exercises. Detailed answers are provided, even to the exercises requiring verbal responses. Solutions to the computational exercises were double-checked using two different sets of statistical software. Step-by-step solutions are provided for many exercises. When they are not, various values obtained at intermediate steps in the solution are shown so that you are able to pinpoint where an error has been made in the event you obtain the incorrect solution.

We have continued the use of the decision-making chart as an appendix. This chart will assist you in deciding which statistical procedures are appropriate for a given purpose. Refer to it frequently. As you gain familiarity with its use, you can overcome a common lament, "I know what to do and how to do it but I am not always sure when to do it."

Much of your work may involve the use of a pocket calculator or even a computer. Some suggestions are provided concerning the calculator characteristics that are likely to be most useful to you in this course. In addition, there also follows a section that provides sources of inexpensive or cost-free statistical software.

It should be emphasized that the Study Guide is not intended as a substitute for the text but, rather, to serve as a supplement. If used in conjunction with the text, as recommended, the Study Guide should enhance your comprehension and enjoyment of the course.

A word of thanks to David Cuevas of El Paso Community College for his review of the manuscript for this edition.

Richard P. Runyon
Audrey Haber

Special Note to the Student

If you are presently in the market for a pocket calculator, we recommend that you purchase one that includes the following functions.

+	Used to add one quantity to another.
−	Used to subtract one quantity from another.
×	Used to multiply one quantity by another.
÷	Used to divide one quantity by another.
X^2	Used to obtain the square of a number.
$\sqrt{}$	Used to obtain the square root of a number.
CLR (or C)	Used to clear all information in the calculator.
CE	Used to clear your last entry in the event you entered the wrong value and wish to remove it without erasing the previous information in the calculator.
MS	Used to place a quantity in storage for future use.
MR	Used to recall a quantity that was previously placed in memory store.
M+	Used to add one quantity to another already in memory store. Excellent for cumulating the squares of numbers.
M−	Used to subtract a quantity from another quantity already in memory store.

Most calculators, although not all, are designed to follow a natural flow in the solution to arithmetic problems. Thus, to add three numbers (e.g., 5, 8, 16), you would follow this sequence (reading from left to right, left to right, etc.):

Value of variable	Function	What you see in display	
5	+	5	
8	+	13	
16	=	29	Answer

The sequence reads as follows: enter 5, enter +, enter 8, enter +, enter 16, enter =. The last number in the display provides the answer.

To add and subtract a series of numbers, such as $15 - 8 + 23 - 14$, you would observe the following sequence:

Value of variable	Function	What you see in display	
15	−	15	
8	+	7	
23	−	30	
14	=	16	Answer

To square and sum a series of numbers, for example, $10^2 + 5^2 + 3^2$, the following sequence is usually employed:

Value of variable	Function	What you see in display	
10	X^2	100	
	+	100	
5	X^2	25	
	+	125	
3	X^2	9	
	+	134	Answer

If your calculator does not contain the function X^2, you may still solve this problem in the following way:

Value of variable	Function	What you see in display	
10	×	10	
10	+	100	
5	×	5	
5	+	125	
3	×	3	
3	+	134	Answer

This is but a brief introduction to the use of the pocket calculator. If you purchase one—alone or with friends—we recommend that you spend a few hours with the manual of operating instructions. Total familiarity with the capabilities of your calculator will save you much time and some grief.

Some Sources of Inexpensive or Cost-Free Statistical Software

Those of you who have invested in commercially available statistical software, of which such packages may easily run into the hundreds of dollars, occasionally, will find, as I have, that some of the programs are seriously flawed. If you are not already aware, it should come as a pleasant surprise that many of our colleagues—most in college and university settings—have developed programs for use with the PC and offer them at no cost to whoever requests the program listings. In some cases, university policy does not permit the reproduction costs nor the costs of mailing and handling to be borne by the department. In such cases, a small fee is charged to help offset these costs. I have indicated where such fees have been requested. **The banks of many countries charge a high fee to cash checks that are drawn on banks in the United States. For example, in Canada, it costs $2 to cash a check from the United States regardless of the size of the check.** If you enclose funds to help cover reproduction, mailing, and handling costs, please send a Postal Money Order. If you are into computers, the following sources of listings in statistics may be useful to you.

Some of these programs may require modification to make them compatible with your computer. I use the Radio Shack TRS-80, Model II and the TRS-80 Color Computer II. With rare exception, I have experienced little difficulty in adapting the programs for my use.

Generalized kappa coefficient: A Microsoft BASIC program. For use with IBM PC and Apple II microcomputers. Measures nominal rating agreement among observers or raters. Allows up to 500 subjects, 10 observers, and 15 categories. Write to: J. Oud, Katholieke Universiteit, Institute voor Orthopedagogiek, Erasmusplein 1, 6500 HD Nijmegan, The Netherlands, or J. M. Sattler, Dept. of Psychology, San Diego State U., CA 92182.

The following listings may be obtained from: Alfred L. Brophy, Ph.D., Behavioral Science Associates, P.O. Box 748, West Chester, PA 19381. Enclose stamped, self-addressed envelope for each program listing.

Approximation of the inverse normal distribution function. Shows and compares eight approximations of the inverse of the normal distribution function, in BASIC. Input probability value and obtain corresponding z. Written in TRS-80 Level II BASIC.

Ranking programs for matched groups and combined distributions. Sorts and ranks N scores of each of M-matched groups. Written in TRS-80 Level II BASIC.

A BASIC program for Tukey's multiple comparison procedure. Uses Tukey's HSD test for making pairwise multiple comparisons among means of groups with homogeneous variance. Printout displays observed difference between means, q, and the 95% confidence limits. Accuracy and speed of seven approximations of the normal distribution function. Shows and compares seven methods for estimating the probability of the normal deviate, z. Easily modified to also yield percentile ranks. Program written for TRS-80 Model I. Took only a few minutes to enter on my TRS-80 Level II BASIC.

A versatile sorting and ranking program. Sorts and ranks scores and produces two types of output: (1) scores in sorted order, along with associated ranks and subject identification numbers; (2) ranks of the scores in their original order. Runs on TRS-80 Models I, II, and III. Readily adaptable to IBM PC and Apple II.

The following listing packages may be obtained from: Daniel Coulombe, University of Ottawa, 275 Nicholas, Ottawa, Ontario, Canada K1N 6N5.

Professor Coulombe offers a marvelous series on one-way ANOVA and two-way ANOVA with and without repeated measures, tests of simple main effects, and multiple comparisons for microcomputers. All the listings are available for a single fee of $5.00 to cover production costs, mailing, and handling. Request ANOVA package and make Postal Money Order payable to his name.

Professor Coulombe also offers a series of multivariate programs—principal component analysis, multiple regression analysis, multiple discriminant analysis, two-way multivariate analysis of variance, correlation matrix analysis, and utilities. All the listings are available for a single fee of $10.00 to cover production costs, mailing, and handling. Request multivariate package and make Postal Money Order payable to his name.

Program for chi square and normal curve integration. Integrates the t density function to provide two-tailed t probabilities with 5-decimal-place accuracy, up to 115 df. Also determines chi square and normal curve probabilities. The program ran flawlessly the first time I entered it on disk. Written in Microsoft BASIC. For program listing, write: D. Louis Wood, Ph.D., University of Arkansas at Little Rock, 33d and University, Little Rock, AR 72204.

PSYCHO-STATS 80: A basic statistical package for the TRS-80, 32K Model I. Includes many descriptive and inferential statistical analyses, including one-, two-, and three-way ANOVAs, multiple and pairwise comparisons, several analyses of covariance, correlational analyses, and various nonparametric tests of significance. Programs are on three 5-inch diskettes. Write to: David E. Anderson, Allegheny College, Meadville, PA 16335. Make check of $20 (to cover cost of diskettes) payable to the Department of Psychology, Allegheny College.

In addition to the above, *Behavior Research Methods and Instrumentation* carries an enormous variety of statistical programs in each issue of this fine journal. In many cases, the program listings actually appear in the article. In some instances, however, you must write to the author(s) to obtain program listings.

STUDY GUIDE
AND SOLUTIONS MANUAL
to Accompany

FUNDAMENTALS
OF BEHAVIORAL
STATISTICS

PART I

INTRODUCTION

The Definition of Statistical Analysis

BEHAVIORAL OBJECTIVES

1. State the two functions of statistical analysis and provide examples of each.

2. Define the following and give examples that are not included in the text: variable, data, population, parameter, sample, and statistic.

3. Describe the purpose of and distinction between descriptive and inferential statistics. Cite examples of each.

4. State the three broad categories in which the goals of research may be classified.

5. Distinguish between a *true experiment* and an experiment that masks as one. Define and distinguish among *independent, organismic,* and *dependent* variables.

CHAPTER REVIEW

To many people, statistics is a collection of numerical facts about the world in which we live and the people who populate this world. Statements such as the following characterize this view: Only three out of ten people know the history of the city in which they live; in a career that spanned 23 years, Hank Aaron blasted 755 home runs; Rita runs the mile in under 6 minutes; at 1454 feet in height, the Sears tower in Chicago is the world's tallest building.

In contrast, most scientists regard statistics as a method of dealing with data. It involves the organization and analysis of numerical facts or observations that are collected in accordance with a systematic plan. You, as students of statistics, are expected to learn some of the fundamental techniques and procedures for gathering, organizing, and performing analyses of data that are collected in your field of study.

A scientist who spends years studying, collecting data, and eventually arriving at a number of conclusions concerning the alcoholic in American society is likely to regard the field of statistics as an integrated process for dealing with data rather than as a mere collection of numerical facts.

If we regard statistics as a method of dealing with data, we can subdivide statistics into two types—descriptive and inferential statistics. Thus, the behavioral scientist using statistics as a method of handling data might have two purposes in mind: first, to describe data and then to explore certain hypotheses in which the interpretation of the data occupies a central role. Typically, when exercising this inferential role, the scientist wishes to draw broad conclusions about large groups of data that are based on smaller samples taken from these large groups.

If the behavioral scientist were simply to draw up a table providing family background information (number of brothers and sisters, history of divorces in the immediate family etc.) about the alcoholics in the study, he or she would be using statistics in a descriptive manner. However, if he or she were to investigate the hypothesis that alcoholics have a greater incidence of alcoholism in their family history than nonalcoholics, he or she would be using inferential statistics. In short, statistics used to present an overall picture of the sample data are considered descriptive. If, on the other hand, the data are used as a springboard to formulate conclusions about the population from which the sample data were drawn, the scientist would be exercising the inferential function.

To understand fully the preceding statement, you should know the meaning of the word "population" as it is used here. A *population* is defined as a complete set of individuals, objects, or measurements having some common observable characteristics. In the example of the scientist studying alcoholism, the population is all the individuals who can be classified as alcoholic.

A *sample* is a subset of a population. Those alcoholics who were chosen to be included in the study were a sample from the population of alcoholics. In other words, the sample consisted of a small group of individuals who were selected from the larger group, or population, of alcoholics. If we wish to make comparisons with the population of nonalcoholics, we would also select a sample of nonalcoholic individuals for study. Data are collected from both groups of sample subjects, descriptive statistics are calculated for each group (e.g., the proportion of alcoholics versus nonalcoholics with alcoholism in their immediate family history), and then inferential statistics are enlisted in order to draw conclusions concerning possible differences in the alcoholic family history of both populations.

Let's take a look at an actual example. Among the many activities that have made the Center for Disease Control (CDC) the focus of worldwide attention and acclaim are the many statistical files it gathers on various factors that influence the health, safety, and well-being of people. In addition to its well-known and publicized epidemiological studies, the CDC initiated, in 1984, a series of surveys of various behaviors that place people at risk for injury, illness, and mortality.

At present, 36 states participate in the behavioral risk-factor surveillance (BRFS). Included in the surveys are such risk factors as smoking, overweight, sedentary life-style, binge drinking, and seat belt nonuse, to name a few. Using a survey developed jointly by the states and CDC, interviewers reach samples from the adult population via a random-digit-dialing telephone technique and solicit data on a number of behaviors that are known to place the individual at risk for injury or disease. Figure 1.1 presents a bar graph that shows current smoking prevalence in three different age groups and three different educational classifications (*Morbidity and Mortality Weekly Report,* **38**, no. 49, 845–848). The various bars in the figure are based on descriptive statistics calculated from the sample survey results. From an inspection of the bar graph, two trends appear that warrant further investigation and the application of inferential statistical techniques: (1) in all age groups, educational level and smoking appear to be related—the greater the education, the lower the rate of smoking; (2) the rate of cigarette smoking among individuals in the older age group (55 years and over) appears to be less than among the younger age groups. In the words of the report (p. 847), "These results indicate that limited progress has been made in reducing the prevalence of cigarette smoking among young adults of low educational attainment levels—a finding consistent with other surveys."

A Few Words About Research Design

As noted in the text, the research design outlines the strategy for collecting data in such a way that specific questions can be answered and/or evidence obtained that permit the confirmation or disconfirmation of research hypotheses. Although this sounds simple enough, it can be and often is a thorn in the side of the researcher. Take the mail survey reported in Case Example 1.1 Ideally, we want the sample to be a mirror image, in miniature, of the population to which we wish to generalize our findings. But there are many psychotherapists out there. We want this sample to include a representation of most, if not all, the therapies in present use (over 130 by a count by Parloff in 1976). And certainly, we want the sample to consist of male and female therapists who are reasonably representative with respect to geographical, ethnic, religious, racial, and other relevant variables.

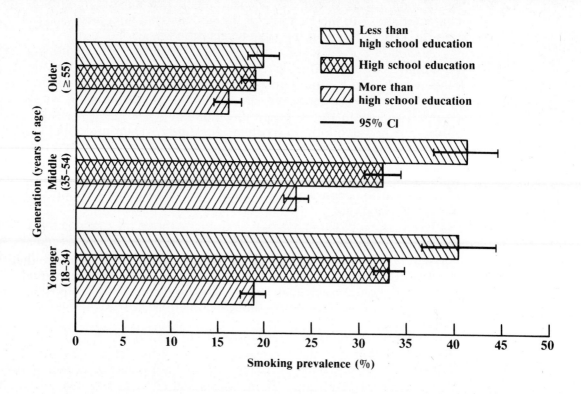

Figure 1.1   **Prevalence of current cigarette smoking, by generation and educational attainment (Behavioral Risk Factor Surveillance System, 1988).**

Then there are other problems. In a mail survey, many people selected in the sample do not respond, either because of the bother of answering the questions or for fear of revealing something unsavory about themselves. There is also the possibility that, among those who do respond, some will distort answers to put themselves in a good light.

How did Pope and his colleagues overcome these manifold problems? To reduce somewhat the irritation of responding to a questionnaire, they included a return envelope with each questionnaire. Forms were randomly sent to 1000 (500 men and 500 women) of the 4456 members of Division 42 of the American Psychological Association (Psychologists in Private Practice). Random sampling is like drawing out of a hat—each member of the population is as likely to be selected as any other member. Incidentally, 585 (58.5%) responded, which is quite high for a mail survey Moreover, the final sample represents about 13% of the membership (585/4454 · 100). Finally, to discourage falsification of responses, all questionnaires were answered anonymously.

The problems of designing a good survey are quite different from those involved in designing a good experiment The purpose of a survey is usually to obtain some estimate of a parameter (e.g., the proportion of psychotherapists of either gender who have been attracted to their patients).

In contrast, the typical experiment attempts to assess the effects on a dependent measure of manipulating an independent variable. The problem here is that many true experiments simply cannot be conducted because of ethical, logistic, and/or economic considerations. This is particularly the case for the caregiving professions, such as dentistry, nursing, medicine, psychotherapy, and so forth. Who would seriously suggest that we deliberately inflict lesions in the brains of human subjects so that we can establish a causal relationship between the lesioned areas and human performance? Rather, we should wait until a head trauma occurs accidentally and then investigate the effects after-the-fact. The research design has the superficial appearance of an experiment. There are usually two conditions—the experimental group (victims of a lesion in a specific area of the brain) and a control group (often individuals judged to be similar to the experimental subjects but with no brain trauma or, better still, individuals who have undergone brain trauma but with injury to a different part of the brain). The dependent measures are typically different aspects of human performance—verbal fluency, arithmetic ability, reaction time, and so on.

The problem with these studies that masquerade as experiments (commonly referred to as pseudo-experiments) is that the results are often interpreted as reflecting causality. Apparent changes in performance measures of the experimental group may be misinterpreted as having been caused by the injury to a specific brain location. However, we must not overlook the possibility that the experimental subjects were different to begin with. Perhaps some of them lacked the ability to appraise common dangers accurately and, as a result, they are greater risk-takers than most of their peers. Or perhaps those involved in vehicular accidents are more likely to have been on drugs (alcohol or cocaine, e.g) and that the drugs may have previously inflicted brain damage that impaired skilled performance. Whenever subjects are assigned to experimental conditions on the basis of preexisting differences, this ambiguity in interpreting the results prevails. Generally, we lump together these preexisting conditions and refer to them, rather unsatisfactorily, as *organismic variables*. Thus, gender, intelligence quotient, ethnic background, and emotional stability are organismic variables, as are acquired characteristic features of the individual's makeup, such as heavy smoker versus light smoker versus no smoker, drug abuser versus nonabuser, vegetarian versus omnivore, anorexic versus obese, and so on. The point is that when we compare some aspect of behavior (dependent measures) of smokers versus nonsmokers, drug abusers versus nonabusers, vegetarians versus omnivores, or anorexic versus obese individuals, any observed differences on the dependent variables may not be due to differences in the so-called independent variables by which they were classified. Let's contrast a few examples of pseudo-experiments with true experiments. Note that, in all the examples shown in Table 1.1, a true experiment is unlikely to be performed because of ethical, logistic, and/or economic considerations.

As you can see, on ethical grounds alone none of these true experiments would be performed on human subjects. However, ethical concerns notwithstanding, how could we adequately monitor and control feeding and martial stress in the last two examples given? As you can see, the various caregiving professions often have no choice about the designs of many of their studies. They are forced to rely on pseudo-experiments for data on the relationship between independent and dependent variables. The problem comes when causal significance is conferred on these relationships. This is often done by the members of the media among whom the intricacies of inference making in health and life science research are commonly misunderstood. Thus, the media may report that obesity *causes* heart attacks, whereas research has only shown that the two *are related*. Obese people may have a greater rate of heart attacks for many reasons other than weight: due to life-style differences (they may exercise less) or due to adverse underlying physiological factors that produced their obesity to begin with.

Incidentally, pseudo-experiments based on independent and dependent variables collected after-the-fact are referred to as *retrospective studies*. Thus, if we went to past records to identify obese and nonobese individuals and recorded the rate of heart attacks in both groups, the study would be retrospective. When the independent measures are based on already existing data but the collection of the dependent data is extended into the future, the research strategy is known as a *prospective study*. Thus, we might identify individuals who are presently obese or nonobese and then follow their future health records to learn if their rates of heart attacks differ.

Table 1.1   Examples in Which a True Experiment Will Not Be Performed Because of Ethical, Logistic, or Economic Reasons

True experiment	Pseudo-experiment
Purpose: To determine the effects of smoking (independent variable) on various aspects of health and/or behavior (dependent variables)	
Smoking conditions assigned to experimental subjects, usually on some random basis. Control subjects are assigned to nonsmoking conditions.	Subjects who are already smokers become experimental subjects. Nonsmokers become control subjects.
Purpose: to determine the effects of an illicit drug (independent variable) on cognitive tasks, psychomotor performance, and/or health (dependent variables)	
Illicit drugs are assigned to experimental subjects. Control subjects are assigned to nondrug conditions.	Subjects who are already illicit drug users become experimental subjects. Nonusers become control subjects.
Purpose: to determine the effects of obesity (independent variable) on the rate of heart attacks (dependent variable)	
Experimental subjects are forced to become obese, perhaps by force-feeding. Control subjects are prevented from becoming obese.	Heart attack rates of subjects who are already obese are compared to rates of those of normal weight.
Purpose: to determine the effects of marital stress (independent variable) on various measures of personality, mood, and adjustment (dependent variables)	
Experimental subjects are purposely exposed to stressful marital conditions. Control subjects are not allowed to experience marital stress.	Dependent measures of subjects who are undergoing or who have undergone marital stress are compared to subjects who are not undergoing marital stress.

SELF-QUIZ: TRUE-FALSE

True/False exercises will be included with every chapter of this Study Guide. After you have completed and scored each quiz, focus your attention on the items where you guessed correctly or answered incorrectly. Try to state, in your own words, the reason that a given answer is correct.

Circle T or F.

T F 1. Statistics is concerned only with the collection of numerical facts.

T F 2. A variable is a theoretical or complete set of individuals, objects, or measurements that have some common observable characteristic.

T F 3. A parameter is any characteristic of a sample that is measurable.

T F 4. A constant is frequently, although not always, a variable.

T F 5. The research design is the plan for collecting data.

T F 6. The time required to complete a task is likely to be a dependent variable.

T F 7. Inductive statistics is concerned with making inferences about populations that are based on samples taken from populations.

T F 8. A population is a theoretical or complete set of objects, individuals, or measurements that have some common observable characteristic.

T F 9. Parameters are almost always known.

T F 10. A variable consists of numbers or measurements collected as a result of observations.

T F 11. We often use a statistic that is calculated from a sample to estimate a parameter.

T F 12. Of all the children age 5 in the United States at a particular time, the proportion who are female is a parameter.

T F 13. To estimate the number of defectives in the daily output of a production process, 300 items are selected and tested. The 300 items constitute a population.

T F 14. Statistics has few applications to daily life.

T F 15. Greek letters indicate statistics rather than parameters.

T F 16. Fundamentally, the purpose of calculating statistics from a sample is to understand the characteristics of that sample.

T F 17. All men who wear glasses is a subset of the population of men.

T F 18. Tables, graphs, and figures are examples of inferential statistics.

T F 19. In most studies, inferential statistics follows descriptive statistics and not vice versa.

T F 20. The design phase of research should take only a fraction of the total time that is allocated for the research.

T F 21. In random sampling, each member of a population does not have an equal chance of being selected.

T F 22. Dosage level, time deprived of food, and gender may all serve as dependent variables in a study.

T F 23. Data are a direct result of observations.

T F 24. The only acceptable definition of statistics is that it is a method for dealing with data.

T F 25. It is not possible to lie with statistics.

T F 26. A census is a random sample that is selected from a large population.

T F 27. In inferential statistics, the investigator can explore hypotheses that he or she holds about the general population.

T F 28. In the case example dealing with the sexual attraction to clients in psychotherapy, the dependent measures were the responses to the questionnaire.

T F 29. An organismic variable is a state of nature that is not under the experimenter's control.

T F 30. In the study on seasonal affective disorder, the score on a mood survey was the independent variable.

T F 31. When an organismic variable is used in a study as the independent variable it is usually possible to establish a direct causal relationship between the independent and dependent variables.

Answers: (1) F; (2) F; (3) F; (4) F; (5) T;(6) T; (7) T; (8) T; (9) F; (10) F; (11) T; (12) T; (13) F; (14) F; (15) F; (16) F; (17) T; (18) F; (19) T; (20) F; (21) F; (22) F; (23) T; (24) F; (25) F; (26) F; (27) T; (28) T; (29) T; (30) F; (31) F.

SELF-TEST: MULTIPLE CHOICE

Multiple-choice exercises will be included with every chapter of this Study Guide. After you have completed and scored each quiz, focus your attention on the items where you guessed correctly or answered incorrectly. Try to state, in your own words, the reason that a given answer is correct.

1. A characteristic or phenomenon that may take on different values is referred to as a:

   a) constant   b) data   c) population   d) variable   e) parameter

2. The plan for collecting data is called the:

   a) descriptive function   b) inferential function   c) research design
   d) statistical analysis e) random sampling

3. A characteristic of a population that is measurable is a:

   a) statistic   b) sample   c) parameter   d) datum   e) constant

4. A number resulting from the manipulation of raw data according to certain specified procedures is called a:

   a) statistic   b) parameter   c) sample   d) population   e) constant

5. In order to estimate the ratio of males to females in a college, a professor decides to calculate the proportion of males and females in a class. The resulting proportion is a:

   a) statistic   b) parameter   c) constant   d) population   e) sample

6. The proportion of all male to all female adults of voting age in the United States at a given time is a:

   a) statistic   b) parameter   c) constant   d) population   e) sample

7. To study the effects of food deprivation on activity, subjects are deprived of food for a period of 10 hours. The time of deprivation is a(n):

   a) independent variable   b) organismic variable   c) dependent variable
   d) datum   e) descriptive statistic

8. In Exercise 7, the activity measure is a(n):

   a) independent variable   b) organismic variable   c) dependent variable
   d) datum   e) descriptive statistic

9. In inferential statistics, the purpose is often to make inferences about _____ that are based on _____ taken from the _____.

   a) samples, populations, samples
   b) statistics, samples, populations
   c) statistics, populations, samples
   d) populations, samples, populations
   e) samples, statistics, populations

10. When a television rating service reports that about 25 million people viewed a particular program, the statement represents a:

   a) wild guess   b) sample statistic   c) descriptive statistic
   d) population parameter   e) statistical inference

11. In a complete census of a suburban community, it was found that 53% of the families have two or more children. The 53% represents a:

    a) parameter   b) population statistic   c) sample parameter
    d) statistical inference   e) constant

12. Which of the following is the correct usage of the term "statistics"?

    a) Martha bench-pressed 325 pounds
    b) speaking of sales statistics, we sold a coat worth $25,000 last week
    c) at Ybrik University, Ray is just one more statistic
    d) the annual per capita cost of educating a student at Ybrik University is $10,000
    e) all of the above

13. When a research team finds that drug use and lowered academic motivation appear to go together in its study, it is:

    a) describing a relationship
    b) demonstrating that drug abuse causes lower academic motivation
    c) reporting a parameter
    d) engaging in speculative thinking
    e) none of the above

14. In Exercise 13, the independent variable is:

    a) unknown   b) academic motivation   c) the research findings
    d) drug use   e) the research design

15. In statistics, Greek letters are commonly used to represent:

    a) samples   b) statistics   c) data   d) parameters   e) variables

16. In performing the descriptive function, behavioral scientists:

    a) make broad inferences about populations, based on sample statistics
    b) determine whether or not there has been an effect of the independent variable on experimental subjects
    c) focus their inquiry on estimating values of parameters
    d) all of the above
    e) none of the above

17. Measurement of four individuals reveals their weights to be, in pounds, 120, 220, 185, 147. These numbers represent:

    a) data   b) statistics   c) variables   d) parameters   e) samples

18. A scientist investigating the effects of environmental temperature on the rate of chirping in crickets must draw the sample from the population of:

    a) all chirping insects
    b) arctic and tropical insects
    c) all crickets
    d) all-weather data
    e) different sounds that crickets make

19. In Exercise 18, the independent variable is:

    a) crickets
    b) rate of chirping
    c) all sounds made by crickets
    d) the time of the year
    e) environmental temperature

20. In Exercise 18, the dependent variable is:

    a) crickets
    b) chirping sounds
    c) all sounds made by crickets
    d) the time of the year
    e) environmental temperature

21. In performing the inferential function, behavioral scientists:

    a) make broad inferences about populations, based on sample statistics
    b) determine whether or not there has been an effect of the independent variable on experimental subjects
    c) focus their inquiry on estimating values of parameters
    d) all of the above
    e) none of the above

22. Which of the following is *not* a variable?

    a) a person's weight
    b) the number of inches in a mile
    c) Janet's blood pressure
    d) the gender of a random sample of subjects
    e) daily rainfall

23. The symbol for the proportion of all cars stolen last year in the United States should be represented by:

    a) a black slash mark   b) a letter in italics   c) a numeric coefficient
    d) an exponent   e) a Greek letter

24. Which of the following is *not* an organismic variable?

    a) gender
    b) prior history of smoking
    c) eye color
    d) pulse rate
    e) time since last meal was consumed

25. Which of the following is *not* likely to be an independent variable?

    a) reaction time
    b) the *amount of reward* following a correct response
    c) the *noise level* in an experimental setting
    d) the amount of light exposure of SAD subjects
    e) the type of psychotherapy used in treatment

Answers: (1) d; (2) c; (3) c; (4) a; (5) a; (6) b; (7) a; (8) c; (9) d; (10) e; (11) a; (12) d; (13) a; (14) d; (15) d; (16) e; (17) a; (18) c; (19) e; (20) b; (21) d; (22) b; (23) e; (24) e; (25) a.

Basic Mathematical Concepts

BEHAVIORAL OBJECTIVES

1. Define mathematical nouns, adjectives, verbs, and adverbs, and identify examples of each.

2. State the summation rules and the various generalizations that are based on these rules.

3. Distinguish among the three types of numbers, citing examples of each. Describe the three ways in which numbers can be used.

4. Define and distinguish among the nominal, ordinal, interval, and ratio scales of measurement. Generate examples that illustrate the distinctions among these scales.

5. Describe the difference between discrete and continuous measurement scales. Explain the types of errors that are commonly associated with each.

6. List the rules of rounding and apply them to real sets of data.

7. Distinguish among ratios, frequencies, proportions, and percentages. Convert proportions to percentages and vice versa.

8. Calculate proportions and percentages of frequency data that are classified into two nominal categories down the columns, across the rows, and in terms of totals. Interpret the resulting proportions and/or percentages and explain or resolve the apparently contradictory results.

CHAPTER REVIEW

To communicate, humans make use of symbols. Everything around us can assume meaning beyond its mere presence. A slab of granite can become a tombstone, a number can become a quantity. The number and the granite are symbols.

Just as we must know words and the rules of their use (their grammar) in order to communicate thoughts, so also we must know the symbols and the grammar of mathematics in order to communicate statistical concepts. Like written and spoken language, mathematics has nouns, adjectives, verbs, and adverbs. For example, "X" and "Y" are often used in mathematics as nouns, representing quantity, scores, or values of a variable. Suppose you are conducting a study in which the two variables are body temperature and the hours since the last food intake. You might identify these variables as X and Y, respectively. These symbols would be mathematical nouns.

Another noun often used in statistics is "N," which stands for the number of subjects, scores, or quantities with which you are dealing. If you had the following scores—11, 707, 21, and 88—the N would be equal to 4. Thus, X, Y, and N are examples of mathematical nouns.

Subscripts are mathematical adjectives that give more precise information about the nouns they modify. In the series, X_a, Y_8, and N_2, the subscripts a, 8, and 2 are adjectives modifying the nouns, X, Y, and N.

Just as the word "grow" is a verb in English, so also are the symbols $+$, $\sqrt{}$, Σ, and $\div$ verbs in the world of statistics. They instruct the reader to perform specific operations, such as add, extract the square root, sum a series of quantities, and divide.

Finally, there are adverbs in statistics that modify the verb, as in written and spoken language. In the notation

$$\sum_{i=3}^{10} Y_i$$

Y is a noun, $\sum$ is a verb, i is an adjective, and $i = 3$ and 10 are adverbs that modify the verb $\sum$ and tell us to sum the third through the tenth values of Y.

The proper statistical notation for the expression $(X_7 + X_8 + X_9)$ is

$$\sum_{i=7}^{9} Y_i$$

Numbers are commonly used in one of three different ways: to name or identify, to represent position in an ordered series, or to represent quantity (how much).

When a number is used to name or identify, the category of numbers to which it belongs is referred to as a nominal scale of measurement. Examples include your social security number, post office ZIP, and your uniform number if you engage in organized sports. The intent of a nominal scale of measurement is to identify and classify individuals or objects that have some characteristic in common. If you are a football player, your uniform number identifies you uniquely, but it also identifies your position—backfield, interior line, linebacker, end. All individuals with a common ZIP number live within the limits of a specific geographical region.

The classifications within an ordinal scale indicate the position in an ordered series. Ranks are commonly assigned to individuals and/or objects to locate their relative position in an ordinal scale. Thus, the highest-ranking violinist in a symphony orchestra occupies the first chair, whereas the seventh-ranking violinist occupies the seventh chair.

The highest scales of measurement, the interval and ratio scales, use cardinal numbers. By definition, then, the numerical values that are associated with interval and ratio scales are quantitative, permitting the use of arithmetic operations such as addition, subtraction, multiplication, and division. Except for one major difference, interval and ratio scales are identical. The difference lies in the nature of the zero point. In an interval scale, the value "zero" is arbitrary and does not indicate the absence of the quantity being measured. To illustrate, consider the way in which temperature is usually measured. The units of measurement are equal—one degree Celsius on one part of the scale equals one degree Celsius on another. However, the reading "zero" does not mean the absence of temperature; it is merely the point on the scale at which water freezes at sea level. Similarly, 0°F—while colder than 0°C—is still relatively warm when compared to the point at which there is a total absence of heat, absolute zero. Because of this difference in the location of the zero value, interval scales do not permit the comparison of one quantity to another in terms of ratios—80°C is not twice as warm as 40°C.

There is another way of categorizing quantitative variables, that is, as discrete or continuous. Between values of a discrete variable, there are gaps that contain no values of the variable—commonly, although not necessarily, discrete variables are whole numbers. Consider variables such as the number of transistors in a microchip, family size, or the number of white cells in a blood sample. The number of white cells, for example, can be 50, 124, 213, and so forth. It cannot be intermediate values such as 50.8, 168.9, or 203.6.

On the other hand, all intermediate values can occur with continuous scales of measurement. Time, length, and weight are examples of continuous variables. A value of 3.666666 is just as legitimate as a value of 4 in a continuous measurement scale. A pilot who gives his or her position as 60.7° longitude, 35.3° latitude, and 1515 feet altitude is using three continuous scales.

When reporting values that are derived from a continuous scale, one often needs to round off numbers. For convenience, we often report weight to the nearest pound, height to the nearest inch, altitude to the nearest foot, and latitude to the nearest tenth.

Moreover, when dividing one value of a variable by another, we frequently are left with a remainder that extends far to the right of the decimal place. In the accompanying text, we generally carry all answers to three or more places than were in the original data and round off to two more places than in the original data. Thus, if your original data were in tenths, you would carry your answer to ten-thousandths and round to thousandths.

There are three more concepts that are often used in statistics—frequency, proportion, and percentage. Frequency is no more than a count of items, scores, objects, or whatever is being measured. If 350 coffee beans out of a batch of 10,000 must be discarded as unusable, the frequency of bad beans in this group is 350. Proportion is merely the ratio of the items in one category (such as unusable, defective, etc.) to the total frequency of all items. The proportion of bad beans to the total is 350/10,000 or 0.035.

To obtain the corresponding percentage, you multiply the proportion by 100. Thus, the percentage of bad beans is $0.035 \times 100 = 3.5\%$.

In the behavioral sciences, it is not uncommon to obtain from each subject measurements on two or more variables, which are then displayed in tabular form. To illustrate, Table 2.1 shows the blood alcohol concentrations by age group of 839 unintentional drownings in North Carolina between the years 1980 and 1984 inclusive (*Morbidity and Mortality Weekly Report,* **35**, no. 40, 635–638, 1986).

Table 2.1   Blood Alcohol Concentration (BAC) Below 100 mg% and Equal to or Above 100 mg% Among 839 Unintentional Drowning Victims from Whom BACs Were Obtained

	Blood alcohol concentrations		
Age	Less than 100 mg%	Greater than 100 mg%[*]	Row totals
0 to 14	86	1	87
15 to 29	254	120	374
30 to 44	85	84	169
45 to 59	69	50	119
≥ 60	62	28	90
Column totals	556	283	839

[*]The legal level of intoxication in North Carolina.

Source: Morbidity and Mortality Weekly Report, 1986, **35**, no. 40, 635.

In addition to column and row percentages, there are three sets of percentages we may obtain from these data: percentages in terms of totals, percentages across (i.e., by age grouping), and percentages down (i.e., by blood alcohol concentration).

Using the data in Table 2.1, answer the following questions:

1. In what percentage of drownings was the individual legally intoxicated?

2. In what age group was the greatest percentage of drownings?

3. In what age group was the least percentage of drownings?

4. In what age group was the percentage greatest for BACs at or over the legal limit for intoxication?

5. In what age group was the percentage least for BACs at or over the legal limit for intoxication?

6. Among drowning victims whose BACs were less than 100 mg%, what age group contributed the highest percentage?

7. Among drowning victims whose BACs were at or above the legal limit of intoxication, what age group contributed the highest percentage?

8. Referring back to your answers to 6 and 7, how is it possible for a single age group to contribute the greatest percentage of victims to both the "less than 100 mg group" and the "more than 100 mg group"?

Answers: (1) $283/839 \times 100 = 33.73\%$. (2) Ages 15 to 29: $374/839 \times 100 = 44.58$. (3) 0 to 14: $87/839 \times 100 = 10.37\%$. (4) 30 to 44: $84/169 \times 100 = 49.70\%$. (5) 0 to 14: $1/87 \times 100 = 1.15\%$. (6) 15 to 29: $254/556 \times 100 = 45.68\%$. (7) 15 to 29: $120/283 \times 100 = 42.40\%$. (8) Almost 45% of all unintentional drownings, regardless of state of intoxication, were drawn from the age group 15 to 29. Even if their percentages in the two BAC categories were no different from all other age groups, they would still contribute the greatest percentages of drownings to both categories because of the sheer weight of their numbers.

Discrete Versus Discretized Scales

Ask a few friends their heights and weights, and they will probably answer with whole numbers, such as: 63 inches tall and 128 pounds or 74 inches and 230 pounds. Or, if you ask the outside temperature, you may be told that it is 85°F. On the surface, it might seem that because we tend to express all these values as whole numbers, they must be based on discrete scales of measurement. However, this is not the case. The test of the discreteness or continuity of a scale does not depend on whether or not it is expressed in terms of whole numbers. Discrete scales may occasionally include fractional values. Moreover, we routinely express the values of continuous variables—such as time, height, weight, temperature—as whole numbers rather than as fractional values. It is convenient to do so, and whole numbers often facilitate the communication of numerical data. When we express values of a continuous variable as whole numbers, we may be said to have discretized that variable. However, it must be remembered that the way in which they are expressed does not affect their underlying characteristics.

Ratios, Percentages, Proportions, and Hocus Pocus

Many people who defend our right to bear arms are fond of reciting the fact that it is people, rather than guns, that kill. Guns are merely the innocent instruments through which evil intentions are fulfilled (except for the hundreds killed annually by guns thought to be unloaded). Statisticians also have a favorite expression: Statistics do not lie. It is people who lie. Statistics is merely the innocent instrument of their deceptive intentions (except for the untold hoards who tell a statistical lie out of ignorance). Of all the grounds in which the seeds of statistical deception can be planted, ratios, percentages, and proportions are, perhaps, the most fertile.

In the text, we saw that there can be a legitimate confusion about interpreting enumerative data involving two nominal categories. When we find percentages across, they often appear to be in sharp disagreement with conclusions that are drawn from the same data when percentages are expressed in terms of the column variable or in terms of totals. However, as we saw in the text, this confusion is usually readily dispelled by paying careful attention to the ways in which we express our conclusions. If we are talking about percentages across, we must be sure to precede our statement with a clear indication that we are referring to the percentages within each category of the row variable. The same is true when we express the results of finding percentages in terms of the column variable or in terms of totals.

However, there are many practices in which the confusion does not stem from legitimate causes. Confusion is often the consequence of a deliberate conspiracy to commit hocus pocus with numbers. Among the most culpable are advertising agencies, which are under enormous pressure to sell a product regardless of the merits of the product. Many successful advertising campaigns are based on the premise that the product is more likely to be purchased if it appears to have a broad spectrum of acceptance by users or potential users of the product. A favorite ploy to achieve this objective—which I call the accordion gambit—involves a collapsing of the response categories *after* the survey has been conducted. The use of this ploy is often found when reporting the results of a sensory test (taste, smell, feel, or whatever). Here's how it works.

Half of the subjects are given Brand A (the leading competing brand) to taste followed by Brand B (the advertiser's account). The other half are given the same beverages in the reverse order. The drinks themselves are unidentified and, in a well-conducted study, are unknown to either the subject or the person administering the independent variable. The code identifying the beverages is broken only after all the data are collected. Note that both the subject and the test administrator are *blind* with respect to the beverage administered on each trial. This use of the double-blind technique guards against the possibility that either the subject or the test administrator will slant choices according to personal preferences and/or biases. So far, everything is according to Hoyle.

There are usually three response categories, although there may be more. For purposes of the planned deception, *it is vital that one response category be neutral. It must not indicate any preference of one brand over the other.* Now let's suppose that such a study has been done and the following results were obtained:

Response category	Brand A preferred	No preference	Brand B preferred	Total
Number	204	240	156	600
Percent	34	40	26	100

On the descriptive level, the following statements could legitimately be made: Of the total sample, 34% preferred Brand A, 26% preferred Brand B, and 40% stated no preference. Of those expressing a preference ($N = 360$), about 57% ($204/360 \times 100 = 56.7\%$) preferred Brand A over Brand B and 43% ($156/360 \times 100 = 43.3\%$) preferred Brand B over Brand A. These results are not likely to produce great joy among those charged with the weighty mission of extolling the virtues of Brand B.

But do not worry. There's an easy way out of this embarrassment—simply collapse the three categories into two. One category will remain the same—Brand A preferred—but the other two will be collapsed into a new money-making category—Brand B judged equal to (the no preference category) or better than Brand A. Having used the accordion gambit, let's look at the following table and see how the results stack up:

Response category	Brand A preferred	Brand B better than or equal to Brand A	Total
Number	204	396	600
Percent	34	66	100

It is not difficult to imagine the advertising copy. *In a scientific test of taste preference, two out of three people found Brand B as good as or better than the leading brand. Don't you think it's time you gave Brand B a chance?*

Note that the advertiser did not lie. It merely stated the truth in a form somewhat more congenial to the client. But woe be the day that the ad agency for Brand A gets its hands on the same research data. By collapsing the scale in the opposite direction, the following table is generated:

Response category	Brand A better than or equal to Brand B	Brand B preferred	Total
Number	444	156	600
Percent	74	26	100

Using precisely the same data, arranged in a form more suitable for its purposes, Brand A could claim that people finding Brand A as tasty or tastier than Brand B outnumber those preferring Brand B by a margin of 3 to 1. So why change?

There's a moral to this little tale of deception. Decisions concerning the research design and the form the statistical analysis will take must be made prior to the conduct of the study. You don't collect data first and then ask how to conduct the statistical analysis. To repeat, the design phase of research includes decisions about the details of the data analysis. There should be no departures from these decisions without legitimate cause. If departures from the original plan must be made, they should be duly noted in the research report along with the justification for the changes.

A Worked Example: Finding Percentages in Terms of Totals, the Row Variable, and the Column Variable

Traveler's diarrhea is a disorder that inflicts itself on many sightseers who are visiting lands where sanitary standards are less than those in their homeland. It can incapacitate its victims for many days and make a nightmare out of a sojourn that was preceded by high hopes and keen anticipation. In one study, conducted on 39 Peace Corps volunteers in Kenya, a compound (doxycycline) was administered to 18 volunteers over a period of 3 weeks. The remaining 21 subjects were administered a placebo—a substance that resembled the drug but possessed no medicinal properties. The results of the study are shown in Table 2.2.

Table 2.2 Subjects Contracting and Not Contracting Disorder Among Those Administered a Drug or a Placebo

Experimental condition	Outcome of study		Row totals
	Contracted disorder	Failed to contract disorder	
Drug group	1	17	18
Placebo group	9	12	21
Column totals	10	29	39

1. Find the percentages in terms of totals.

2. Find the percentages in terms of the row variable.

3. Find the percentages in terms of the column variable.

4. Among the subjects who received the drug, what percent contracted the disorder?

5. Among the subjects who did not contract the disorder, what percent were in the drug group?

6. Among the subjects who were administered the placebo, what percent contracted the disorder?

7. In your own words, describe the sample results.

Answers:

(1) In terms of totals:

Experimental condition	Outcome of study		Row totals
	Contracted disorder	Failed to contract disorder	
Drug group	2.6	43.6	46.2
Placebo group	23.1	30.8	53.8*
Column totals	25.6*	74.4	100

*Slight disparity due to rounding.

(2) In terms of rows:

Experimental condition	Outcome of study		Row totals
	Contracted disorder	Failed to contract disorder	
Drug group	5.6	94.4	100
Placebo group	42.9	57.1	100

(3) In terms of columns:

Experimental condition	Outcome of study	
	Contracted disorder	Failed to contract disorder
Drug group	10.0	58.6
Placebo group	90.0	41.4
Column totals	100	100

(4) 5.6%
(5) 58.6%
(6) 42.9%
(7) Subjects in the sample who received the drug had a much lower incidence of the disorder (5.6%) than those who received the placebo (42.91). Also, those who contracted the disorder were primarily drawn from the placebo group (90%). These data are reexamined in Chapter 17 of the text to see if the inference can be legitimately drawn that the drug was effective in curbing traveler's diarrhea.

SELECTED EXERCISES

I. Addition and Subtraction

Obtain the sum of the following:

1. +9	2. −18	3. +2.35	4. $(9-2) + (4-4) - (2-7)$
−7	− 5	+1.16	= _____
+2	+16	−5.13	
−5	+10	+4.33	5. $6.354 - 2.999 - 5.444$
−6	−3	−3.19	= _____

Subtract the second number from the first in each of the following:

6. 0.5467	7. 7.235	8. 26.03	9. 15	10. −16	11. 0.0090
0.2349	−12.245	32.15	−26	−11	−0.9910

II. Multiplication

Calculate the product for each of the following:

12. 16.4	13. 0.051	14. 11.2	15. 44.59	16. −2.25	17. −0.01
8.2	−1.613	11.2	0.06	1.76	−0.01

III. Division

18. $14 \div 4$ 19. $4.56 \div 2.28$ 20. $(7-4) \div 1.96$

21. $15 \div -45$ 22. $0 \div 5.5$ 23. $-12 \div -12$

24. $-8 \div 2.5$ 25. $.0066 \div .0022$ 26. $15.3131 \div 5.5$

IV. Summation Rules

27. Find $\sum_{i=4}^{5} (X_i)$ when $X_1 = 4$, $X_2 = 5$, $X_3 = 6$, $X_4 = 7$, $X_5 = 8$ and $X_N = 9$.

28. a.   Find $\sum_{i=1}^{N} (X_i)$ for the values given in Exercise 27.

   b.   Find $\sum X^2$ for the values given in Exercise 27.

   c.   Find $(\sum X)^2$ for the values given in Exercise 27.

V. Equations Involving Fractions

29. Solve $a = b/c$ for b when $a = 6$ and $c = 5$.

30. Solve $\bar{X} = \sum X/N$ for $\sum X$.

31. Solve Exercise 30 for N.

32. Solve $X = a + by$ when $a = 5$, $b = 1.5$, and $y = 12$.

33. Solve Exercise 32 for b when $a = 5$, $y = 12$, and $X = 23$.

34. Find the value of SS when SS $= \sum X^2 - (X)^2/N$ when $\sum X^2 = 97.4026$, $\sum X = 72.86$, and $N = 85$.

VI. Complex Operations

Perform the indicated operations:

35. P^3 when $P = 0.05$ 36. $28/\sqrt{16}$ 37. 5^0 38. $9\sqrt{144}$ 39. $4 \div 5^2$

40. $5P^2Q^0$ when $P = 0.5$ and $Q = 0.5$ 41. $(6 - 5 + 7 - 2)/(-6/9)$

42. $5^6/5$ 43. $4^4/4^3$ 44. $8 \div \sqrt{0.0081}$ 45. $9 + \sqrt{36}$ 46. $0.9^2 - 0.5^3$

47. $\sqrt{16 \div 49}$ 48. $5 - 2(4 + 5)(12 - 8) + 5/25$ 49. $9^8 \div 9^6$

50. $4^3 + 5^4(2^3)$

Answers: (1) –7; (2) 0; (3) –0.48; (4) 12; (5) –2.089; (6) .3118; (7) 19.48; (8) –6.12; (9) 41; (10) –5; (11) 1; (12) 134.48; (13) –0.0823; (14) 125.44; (15) 2.6754; (16) –3.96; (17) 0.0001; (18) 3.5; (19) 2; (20) 1.5306; (21) –0.3333; (22) 0; (23) 1; (24) –3.2; (25) 3; (26) 2.7842; (27) 15; (28a) 39; (28b) 271; (28c) 1521; (29) 30; (30) $\sum X = N\bar{X}$; (31) $N = \sum X/\bar{X}$; (32) 23; (33) 1.5; (34) 34.9487; (35) 0.000125; (36) 7; (37) 1; (38) 108; (39) 0.16; (40) 1.25; (41) –9; (42) 3125; (43) 4; (44) 88.89; (45) 15; (46) 6.48; (47) 0.5714285; (48) –66.8; (49) 81; (50) 5064.

SELF-QUIZ: TRUE-FALSE

Circle T or F.

T F 1. A notation commonly used to represent a quantity is the symbol X.

T F 2. The symbol $\sum$ usually represents a quantity.

T F 3. The symbol N is a mathematical noun that represents the number of scores or quantities with which we are dealing.

T F 4. The expression X^a directs us to raise X to the ath power.

T F 5. The number on a football player's uniform is a cardinal number.

T F 6. Nominal numbers indicate position in an ordered series.

T F 7. A particular observation of a variable is called the value of the variable.

T F 8. The lowest level of measurement is found with nominal scales.

T F 9. The classes in nominal scales represent an ordered series of relationships.

T F 10. The algebra of inequalities applies to ordinal scales.

T F 11. The difference between interval and ratio scales involves the difference between an arbitrary and a true zero point.

T F 12. The Celsius scale is a ratio scale.

T F 13. Observations of discrete variables are always exact as long as the counting procedures are accurate.

T F 14. Nominal scales show a direction of differences to a limited extent.

T F 15. Continuous scales must progress by whole numbers.

T   F   16. The numerical values of continuous variables are always approximate.

T   F   17. If weight is expressed to the nearest pound, a person weighing 165 on an accurate scale really weighs between 164.5 and 165.5.

T   F   18. To two decimal places, the number 99.99501 rounds to 100.00.

T   F   19. The subscript for the quantity X_b helps to identify it.

T   F   20. The difference between interval and ratio scales involves the difference between an arbitrary and a true zero point.

T   F   21. If each subject in a study is classified according to ethnic origin, the scale of measurement is nominal.

T   F   22. Frequency data may be continuous.

T   F   23. The following numbers are the only ones possible in a scale of numbers: 20, 40, 60, 80. Therefore, the scale must be continuous.

T   F   24. Out of 42 trees in a botanical garden, 3 were oaks and 8 were palms. Rounded to the nearest hundredth of a percent, the percentage of palm trees in the total is 19.05.

Answers: (1) T; (2) F; (3) T; (4) T; (5) F; (6) F; (7) T; (8) T; (9) F; (10) T; (11) T; (12) F; (13) T; (14) F; (15) F; (16) T; (17) T; (18) T; (19) T; (20) T; (21) T; (22) F; (23) F; (24) T.

SELF-TEST: MULTIPLE-CHOICE

1. Grouping individuals into low, middle, and high categories implies which type of scale?

   a) nominal   b) ordinal   c) interval   d) ratio   e) continuous

2. Height is measured on what type of scale?

   a) nominal   b) ordinal   (c) interval   d) ratio   (e) discrete

3. The symbol N, representing the number of observations, is a mathematical:

   a) verb   b) adjective   c) adverb   d) noun   e) predicate

4. The summation sign Σ is a mathematical:

   a) verb   b) adjective   c) adverb   d) noun   e) predicate

5. The scale characterized by the classification of events, objects, or persons into mutually exclusive categories and whose only mathematical relationships are those of equivalence and nonequivalence is called:

   a) nominal   b) ordinal   c) interval   d) ratio   e) equivalency

6. When someone says that Mary appears smarter than Susan, the scale of measurement is:

   a) ratio   b) nominal   c) interval   d) ordinal   e) standard

7. A truly quantitative scale with an arbitrary zero point is called:

   a) interval   b) ordinal   c) standard   d) ratio   e) nominal

8. An example of a nominal scale is:

   a) weight
   b) order of finish in the National Football League
   c) the apparent size of an object

    d) candidates for political office
    e) socioeconomic status

9. The expression $\sum\limits_{i=1}^{N} X_i$ means:

    a) $X_1 + X_2 + X_3$
    b) $X_1 + X_2 + X_3 + \cdots + X_N$
    c) $X_1 : X_2 : X_3$
    d) $X_3 + X_4 + X_5 + \cdots + X_N$
    e) none of the above

10. The lowest level of measurement is:

    a) ratio   b) interval   c) nominal   d) ratio   e) standard

11. Which of the following represents the highest level of measurement?

    a) order of finish in a horse race
    b) male versus female
    c) temperature on Celsius scale
    d) length in inches
    e) selection of most popular instructor

12. The data employed with nominal scales consist of:

    a) relative position in an ordered series   b) scores   c) variables
    d) continuous numbers   e) frequencies

13. The scale of measurement that is characterized by the use of the algebra of inequalities is:

    a) ordinal   b) interval   c) numeral   d) nominal   e) ratio

14. Variables in which measurement is always approximate because they permit an unlimited number of intermediate values are:

    a) nominal   b) discrete   c) ordinal   d) continuous   e) interval

15. What are the true limits of the measurement 12.4 pounds?

    a) 12.4–12.5   b) 12–13   c) 12.35–12.45   d) 11.5–12.5   e) 12.3–12.5

16. The number 15.00500 rounded to the second decimal place is:

    a) 15.00   b) 15.01   c) 16.00   d) 15.005   e) 15.05

17. What number has, as its true limits, 16.55–16.65?

    a) 16.5   b) 16   c) 16.575   d) 16.625   e) none of the above

18. The number 43.54499 rounded to the second decimal place is:

    a) 43.54   b) 43.60   c) 44.00   d) 43.55   e) none of the preceding

19. The number 15.01500 rounded to the second decimal place is:

    a) 15.02   b) 15.01   c) 16   d) 15.005   e) 15.05

20. The data employed with interval or ratio scales are frequently referred to as:

    a) head counts   b) scores   c) ordinal position   d) ranks   e) none of the preceding

21. If $\bar{X} = \sum X/N$, it follows that:

    a) $\sum X = N\bar{X}$ b) $N = \sum X/\bar{X}$ c) $0 = (\sum X/N) - \bar{X}$
    d) all of the above   e) none of the above

22. Y^0 equals:

    a) 1   b) 0   c) $Y - Y$ d) Y/O
    e) must know value of Y in order to answer

Use the following table to answer Exercises 23 to 30.

A deficiency of choline in the diet leads to disorders of the liver, kidneys, and memory in laboratory animals and is suspected as a factor in similar disorders among human infants. Intellectual deficits and short stature are common among villagers in highland communities of Ecuador. Their diets appear to be low in choline. The following table shows a comparison of choline levels in the milk of nursing mothers in Ecuador and Boston, United States.

Milk choline level	Ecuador	Boston
0 to under 100	22	0
100 to under 200	15	3
200 to under 300	15	2
≥ 300	3	6

Source: Based on Zeisel et al., 1982.

23. The proportion of mothers in the total sample who provided choline levels that were less than 100 was:

    a) 33.33   b) 0.3333   c) 83.33   d) 0.8333   e) 0

24. Among mothers providing choline levels more than 300, the proportion from Boston was:

    a) 66.67   b) 33.33   c) 0.6667   d) 0.0909   e) 9.0909

25. Among mothers from Boston, the percentage providing choline levels that were greater than 300 was:

    a) 54.55   b) 66.67   c) 9.09   d) 0.16.67   e) 0.1667

26. Among mothers from Ecuador, the proportion providing choline levels that were greater than 300 was:

    a) 0.8333   b) 83.33   c) 0.0545   d) 0.5455   e) 5.4545

27. Among mothers from Boston, the percentage providing choline levels that were greater than 200 was:

    a) 100   b) 18.18   c) 72.73   d) 32.73   e) 67.67

28. Among mothers providing choline levels that were less than 100, the proportion from Ecuador was:

    a) 0.4000   b) 100   c) 0.3333   d) 40.00   e) 1.0000

29. In the total sample, the proportion of mothers from Boston who provided choline levels that were equal to or greater than 300 was:

    a) 0.9091   b) 0.6667   c) 0.5455   d) 0.0909   e) 0.1667

30. In the total sample, the proportion of mothers from Ecuador who provided choline levels that were less than 100 was:

    a) 0.3333   b) 1.0000   c) 0.4000   d) 0.8333   e) 0.2273

Answers: (1) b; (2) d; (3) d; (4) a; (5) a; (6) d; (7) a; (8) d; (9) b; (10) c; (11) d; (12) e; (13) a; (14) d; (15) c; (16) a; (17) e; (18) a; (19) a; (20) b; (21) d; (22) a; (23) b; (24) c; (25) a; (26) c; (27) c; (28) e; (29) d; (30) a.

PART II

DESCRIPTIVE STATISTICS

3

Frequency Distributions and Graphing Techniques

BEHAVIORAL OBJECTIVES

Conceptual Objectives

1. State the purpose of grouping data that are obtained from a random sample. Define both random samples and mutually exclusive classes. Specify the advantages and disadvantages of grouping data.

2. Distinguish between apparent and true limits of a class. Define the midpoint and width of a class.

3. State the purpose of using cumulative frequency and cumulative percentage distributions. How is each distribution derived from the frequency distribution?

Procedural Objectives

1. Use the five-step procedure described in the text to group frequency data into mutually exclusive classes.

2. Find the width, midpoint, apparent limits, and true limits of a class.

3. Given a frequency distribution, construct cumulative frequency and cumulative percentage distributions.

CHAPTER REVIEW

Ann Johnson is an insurance claims adjuster. At the close of every month, she must wade through the settled claims for the previous 30 days and make a tally of amounts paid by her company in claims. After a few trials, she discovers her work to be much easier, more systematic, and less subject to error if she constructs a table of dollars paid from the largest to the smallest amount. Then she flips through each claim for that month and puts a mark alongside the dollar value corresponding to each claim After going through all the claims, she can count the number of marks corresponding to each dollar amount and record it in the table. When she has done so, she has constructed an ungrouped frequency distribution of scores, in which each score is a dollar amount.

The following is a portion of one of Ann Johnson's ungrouped frequency distributions:

X	f	X	f
$5000	///	$4500	///
$4900		$4400	/
$4800	////	$4300	
$4700	//	$4200	///////
$4600	/	$4100	//

From the ungrouped frequency distribution of scores, the next logical step is to group the scores into classes that include more than one value of the variable of interest. For example, in the preceding table, one might combine $4100 with $4200, $4300 with $4400, and so forth. The frequencies corresponding to each of these values would also be combined. Thus, the class $4100 to $4200 would contain a frequency count of 8(2 + 6). When this collapsing into groups is done for all the values and their corresponding frequencies, we have constructed a grouped frequency distribution.

Some judgment is required in deciding upon the number of classes to use with any given data set. One must strike a balance between using too many classes—thereby failing to achieve the presentational economy of grouping—and too few classes, which results in the loss of much valuable quantitative information. For example, if we were to collapse Ann Johnson's table into one class, the upper apparent limit of the class would be $5000 and the lower limit would be $4100. The corresponding frequency would be 22. As you can see, we would have no way of distinguishing a small award (say, $4100) in this class from a larger award (such as $5000).

The fewer the classes are into which we collapse our frequency distribution, the more of the original information we lose. However, collapsing is practical to some extent if there are many scores and some scores have low frequencies of occurrence. Therefore, the researcher must scrutinize the data to determine the optimal number of classes.

Once the number of classes has been chosen, there are five steps to follow in setting up a grouped frequency distribution of scores.

1. Subtract the lowest value of the variable (score) from the highest value, then add 1 to find the total number of potential scores. Thus, the total number of potential scores = maximum score − minimum score + 1.

2. Determine the number of scores to collapse into each class. To do this, divide the total number of potential scores (step 1) by the number of classes you decided upon. For simplicity and uniformity, 15 classes is commonly used in the text, although it varies with the characteristics of the data. However, most data in the behavioral sciences can be accommodated by 10 to 20 classes while preserving much of the numerical information and achieving economy of presentation. To illustrate, if there are 60 different scores or potential scores and we wanted 15 classes, we would divide 60 by 15 to obtain a class width of $i = 4$. (*Note:* Do not confuse the number of scores with the number of cases or frequencies. There could be 100 frequencies spread out over 5 different score values or 5 frequencies spread out over 100 score values.)

3. Find the upper and lower limits of the lowest class. Naturally, the lowest score in our data set will be the minimum score of the lowest class. To this, add the number of scores in each class (i, found in step 2) minus 1 to obtain the maximum value for the lowest class. For example, if the lowest score in the original data is 12 and $i = 4$, the apparent limits for the lowest class would be 12 to $12 + i − 1$ or 12 to 15.

4. Follow a procedure similar to step 3 to obtain the apparent limits for the remaining classes. The minimum value for the next class is 1 greater than the maximum value for the class below it. So, if the lowest class has an apparent upper limit of 15, the minimum value (or lower apparent limit) of the next class is 16.

5. Each class must have a frequency that is associated with it (the frequency may be zero). Merely add the frequencies of all the scores within a class to find the total frequencies or cases for that class.

Once you have completed step 5, you have constructed a grouped frequency distribution for your data.

The true limits of a class follow the same principle as the true limits of a number. Although the apparent limits of a class may be 12 to 15, the true limits of that class are 11.5 to 15.5. Similarly, the class 205 to 212 has, as its true limits, 204.5 and 212.5.

Suppose Ken Stubler, a social scientist, wishes to conduct a survey in order to find out the potential support for a nationwide consumer organization with the goal of protecting the consumer against fraud and the sale of inferior products. One of his major concerns would be to select his sample in such a way that all samples of the same size would have an equal chance of being selected. If they do, we can say that the samples have been randomly chosen.

After the technique for selecting the sample has been devised, Mr. Stubler might wish to convert the grouped frequency distribution into cumulative frequency and cumulative percentage distributions. In both these distributions, each entry indicates the number or percentage, respectively, of cases falling below the upper real limit of its corresponding class. In other words, the cumulative frequency below the upper real limit of the third class from the bottom consists of the sum of the frequencies for class 1, class 2, and class 3.

A cumulative percentage distribution is calculated by dividing the cumulative frequency corresponding to the upper real limit of each class by the total N of the sample and multiplying by 100. Since the final entry must include all the cases in the data set, the final cumulative frequency must equal N and the final cumulative percentage must equal 100%.

Often, the next step in the handling of data is to represent the findings graphically. When dealing with bar graphs, histograms, and frequency curves, a common practice that is advocated by many statisticians involves the three-quarter rule. According to this rule, the vertical axis should be approximately three-quarters the length of the horizontal axis, with 0.7 to 0.8 being an acceptable range of variation. Thus, if the horizontal axis is 12 units long, the vertical axis should be approximately 9 units in height. This convention reduces the chance that the researcher's personal biases might influence the way in which the data are presented.

Bar graphs, histograms, and frequency curves are all graphing techniques for representing different types of frequency data. For example, a bar graph is the technique that is best suited to nominally or ordinally scaled variables, whereas interval or ratio-scaled variables require either the histogram or the frequency curve. The physical distinction between bar graphs and histograms is not great. They are essentially the same except that in the histogram the bars are placed contiguously so that no separation is visible.

Since histograms represent interval or ratio-scaled variables, we can assume that equal differences along the horizontal axis are equal. However, we cannot make the same assumption when constructing or viewing bar graphs that present nominally or ordinally scaled data.

Although the frequency curve is different in appearance from the histogram, the histogram is easily converted into a frequency curve when we wish to present continuously distributed interval/ratio variables. By placing a dot at the top of each of the vertical bars in a histogram and then connecting the dots with straight lines, you would obtain a frequency curve. The advantage of the frequency curve over the histogram is that two or more frequency curves are more easily compared, visually, than two or more histograms.

Frequency curves are so valuable in handling and comparing data that names have been given to certain forms of frequency curves. The most popular is the bell-shaped normal curve. The normal curve is symmetrical around its central point. If folded along the central axis, both of its sides would be perfectly superimposed on one another. In other words, each side is a mirror image of the opposite side. In contrast, skewed curves do not have a central point around which the other points fall symmetrically. An asymmetrical distribution can be either negatively or positively skewed. A positively skewed distribution would have fewer frequencies along the higher (right) end of the horizontal axis, whereas a negatively skewed distribution would have fewer cases spread along the lower (left) end of the horizontal axis. Graphically, a negatively skewed curve would have a tail to the left and a positively skewed distribution would tail off to the right.

Other Methods of Preparing Grouped Frequency Distributions

There is far from universal agreement concerning the procedures for casting data into the form of grouped frequency distributions. When presenting vital statistics, particularly as they relate to age (age of marriage, death, divorce, or age of contracting a disease) or economic summaries (such as annual earnings per family unit), one commonly represents the lower limit of the lowest class as zero and the upper limit as *just under* the lower limit of the adjacent higher class. Table 3.1 shows the number of cases of rubella (German measles) per 100,000 by age group in the United States in 1983 and 1987.

Table 3.1   Frequency of Rubella per 100,000 in 1983 and 1987

Age group	1983	1987	Total
0 to under 1	127	141	268
1 to under 5	149	895	1044
5 to under 10	102	324	426
10 to under 15	93	688	781
15 to under 20	95	1071	1166
20 to under 25	117	187	304
25 to under 30	83	153	236
30 and over	80	153	233
Total	846	3612	4458

Note that the classes have unequal widths and the highest class (30 and over) is open-ended. Both of these features are common in the reporting of vital statistics. The class with the smallest width (0 to under 1 in Table 3.1) usually involves a category with high relative frequencies or one in which we have a need for more precise data for that class. Since rubella may have serious consequences for children under 1 year of age, we wish to ascertain with precision the incidence of this disorder among these children. The open-ended class (30 and over), on the other hand, is typically one for which we do not require a high degree of precision. It may be a low-risk group or one in which the relative frequency is diminishingly low so that additional classes are unnecessary.

What is of particular note in Table 3.1 is the fact that the boundary values (0, under 1; 1, under 5, and so forth) represent the *true* rather than the apparent limits of each class. Technically, the upper boundary is a fractional value. For example, the upper real limit of the first class is 0.99999+. However, it may be treated as 1.00 without producing any distortion. The important point is that a child is not assigned to the class *1 to under 5* until his or her first birthday.

Another variation in the construction of grouped frequency distributions involves the procedures for specifying the lowest class. In our text, we use the most straightforward method: taking the lowest score as the lower apparent limit of the first class. This method will occasionally lead to a larger grouping error than the use of alternative procedures that are available. In one alternative method, you first decide on the width of the class. Having done so, you select the lowest class so that the lower limit is evenly divisible by the class width. For example, if your lowest value is 56 and the class width is 5, the lower apparent limit of the first interval is 55, since it is evenly divisible by 5 (i.e., there is no remainder). You will note that most of the grouped frequency distributions in the text conform to this rule. For example, for the data in Table 3.1 in the text, the lowest diastolic blood pressure was 51. The width of the class we selected was 3. (See Table 3.3 in the text.) As you can see, when we divide 51 by 3 we obtain 17, with no remainder.

The Advantage of Ordering the Values of the Variable Prior to Grouping

The entire process of assigning scores to the appropriate class is simplified and made more accurate by arranging the values of the variable in order (ascending or descending) prior to making the assignments to each class. The only requirement is that you *count* the number of values on the ordered list that are included within the limits of each class. To illustrate, we'll use the data from Hsu et al. (1981), as shown in Table 3.2. In this study, the purpose was to learn if there is greater instability of chromosomes among 21 patients with thyroid cancer than among six control subjects. For our immediate purposes, we'll combine the values of these two groups. We'll use a total of eight classes. With values ranging from 1 through 18, the class width will be 2.2. The apparent limits of the lowest class, then, will be 1 to 3.1, with the real limits of 0.95 to 3.15.

Table 3.2   Percentage of Cells Showing Chromosomal Instability Among 21 Patients with Thyroid Cancer and 6 Control Subjects

13.9	12.0	17.1	4.0
10.9	13.0	13.4	1.0
9.3	11.3	6.0	5.0
5.2	12.2	14.0	7.0
14.0	5.0	18.0	3.0
2.0	2.0	13.6	3.0
3.2	2.0	5.4	

Source: Hsu et al., 1981.

After deciding on the number of classes and the interval width, we next arrange the values of the variable in order. Now all we have to do is count the number of values in the sorted list that fall between 1 and 3.1. Looking at the sorted list in Table 3.3, we see that there are six. Similarly, we count five values of the variable falling between 3.2 and 5.3. By continuing this count throughout the list, we obtain the grouped frequency distribution shown in Table 3.3

Some students prefer not to work with fractional (decimal) values when constructing a grouped frequency distribution. If you are a member of their ranks, you can get rid of the decimal points in the original values of the variable by multiplying each value by 10 (100 if the data are to two decimal places). Once this is done, all these values are expressed as whole numbers and the values of the variable range between 10 and 180. Instead of a class width of 2.2, you will have a width of 22. At the final step, you reinsert the decimal points. (See Table 3.4.)

Table 3.3   Sorted List of Values of the Variable and the Grouped Frequency Distribution of These Values

Sorted list		Class	f
18 ⎱ 2		16.4 – 18.5	2
17.1 ⎰			
14 ⎱ 0		14.2 – 16.3	0
14			
13.9		12.0 – 14.1	8
13.6 ⎱ 8			
13.4		9.8 – 11.9	2
13			
12.2		7.6 – 9.7	1
12			
11.3 ⎱ 2		5.4 – 7.5	3
10.9 ⎰			
9.3 ⎱ 1		3.2 – 5.3	5
7			
6 ⎱ 3		1.0 – 3.1	6
5.4 ⎰			
5.2			$N = 27$
5			
5 ⎱ 5			
4			
3.2 ⎰			
3			
3			
2 ⎱ 6			
2			
2			
1 ⎰			

Now you reinsert the decimal in the apparent lower and upper limits of each of the classes in the grouped frequency distribution shown in Table 3.4 and obtain the grouped frequency distribution shown in Table 3.3.

The Frequency Curve and the Frequency Polygon

In studying the literature of the behavioral, social, biological, and/or physical sciences, you will often encounter a *frequency polygon* that is a variation of the frequency curve. However, unlike the frequency curve, the polygon is a *closed* figure in which the curve meets the horizontal (X-axis) at both ends of the distribution. In order to convert a frequency curve into a frequency polygon, you simply add one class to each end of the horizontal axis. Since each of these classes has an associated zero frequency, the curve will meet the X-axis at the midpoint of these two added classes. To illustrate, Figure 3.1 shows the modification of the frequency curve in Figure 3.8 of the text that is necessary to convert it into a frequency polygon.

If it is so easy to convert a frequency curve to a frequency polygon, why retain the distinction between these two types of curves? The answer is simple. For many types of data, the frequency polygon is totally inappropriate. Recall that when you add the two classes at each end of the distribution, you assign a zero frequency to each of these classes. But what if you are dealing with data for

Table 3.4 Sorted List of Values of the Variable Multiplied by 10 and the Grouped Frequency Distribution of these Values

Sorted list	Class	f
180	164 –185	2
171		
140	143 –163	0
140		
139	120 –141	8
136		
134	98 –119	2
130		
122	76 – 97	1
120		
113	54 – 77	3
109		
93	32 – 53	5
70		
60	10 – 31	6
54		
52		$N = 27$
50		
50		
40		
32		
30		
30		
20		
20		
20		
10		

which you have purposely restricted your examination to a limited number of classes? Is it valid to assign a zero frequency to the classes that you have deleted from consideration?

To illustrate, suppose you are interested in graphing the incidence of schizophrenic disorders over five-year periods from 1950 to 1985. If you were to add a class at each end of the distribution and assign a zero to the midpoints (1947 and 1988), you would, in effect, be saying that there *was zero incidence of schizophrenia during these two periods,* an obvious absurdity. Similarly, if you are showing age-related data in which your focus is restricted to only certain age groups, it would be equally absurd to add two age classes and assign a zero to the midpoints of these classes. See if you can think up additional examples from fields like economics, biology, and medicine.

It is for these reasons that we focused on the frequency curve in the text. It is a graphic device that is always appropriate for presenting frequency data that are associated with interval- and ratio-scaled variables. In contrast, the frequency polygon is more limited in its potential applications.

Uses of the Cumulative Frequency and Cumulative Percentage Distributions and Graphs

In Chapters 4 and 5, we'll be looking at several different statistics for pinpointing the interpretation of scores in a distribution. For two of these—the percentile

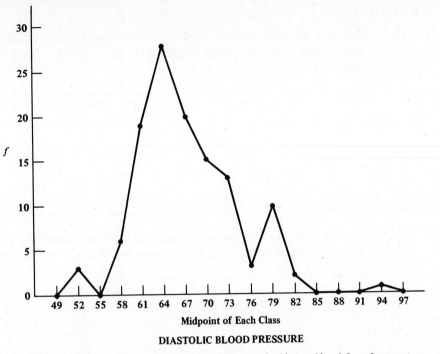

Figure 3.1   Frequency polygon of diastolic blood pressures.

rank of a score and the median—the construction of a cumulative frequency distribution is an essential intermediate step. Moreover, the graph based on the cumulative percentage distribution provides a quick and relatively accurate method of guesstimating the median and various percentile ranks.

To be certain that you understand the construction of cumulative frequency and cumulative percentage curves, fill in the blank spaces in the following table. The chart is a company's tabulation of the monthly income taxes that are withheld from the paychecks of 400 employees.

Real limits of class	f	Cumulative f	Cumulative %	
				17. 400
253.5 to 263.5	2	(17) _____	(16) _____	16. 100
243.5 to 253.5	2	(15) _____	99.50	15. 398
233.5 to 243.5	11	396	(14) _____	14. 99.00
223.5 to 233.5	21	(13) _____	96.25	13. 385
213.5 to 223.5	27	364	(12) _____	12. 91.00
203.5 to 213.5	37	(11) _____	84.25	11. 337
193.5 to 203.5	33	(10) _____	(9) _____	10. 300
183.5 to 193.5	53	267	66.75	9. 75
173.5 to 183.5	54	(8) _____	53.50	8. 214
163.5 to 173.5	44	160	(7) _____	7. 40.00
153.5 to 163.5	31	(6) _____	(5) _____	6. 116
143.5 to 153.5	27	(4) _____	21.25	5. 29.00
133.5 to 143.5	30	58	(3) _____	4. 85.00
123.5 to 133.5	16	(2) _____	7.00	3. 14.50
113.5 to 123.5	7	(1) _____	3.00	2. 28
103.5 to 113.5	5	5	1.25	1. 12
$\sum f = N =$	400			

SELECTED EXERCISES

1. The following are the lowest and highest scores for five different distributions. Fill in the columns with values that best apply. Assume that each distribution has 15 classes.

	Lowest and highest scores	Width of class	Apparent limits of the lowest class	Real limits of the lowest class	Midpoint of the lowest class
a)	9, 23				9
b)	15, 162		15–24		
c)	81, 110				
d)	−10, 52	5			
e)	0, 398	25			

2. The following is a set of hypothetical scores. Group them into frequency distributions employing the following class widths: $i = 1, 3, 5, 8$.

2	35	40	4	43	18	17	28	5
5	19	37	10	39	19	30	24	29
9	16	20	13	17	10	25	19	15
13	17	18	17	12	33	8	7	30
11	42	21	27	21	26	19	22	24
8	37	19	15	13	41	21	35	32
23	12	15	19	8	3	24	19	16
13	19	28	32	26	15	17	21	15
20	34	17	37	14	22	39	20	

Compare and contrast the economy of employing the various widths to the loss of information from the use of progressively wider classes.

3. Construct cumulative frequency and cumulative percentage distributions for each of the frequency distributions in Problem 2.

4. Determine the midpoints of the classes, cumulative frequencies, and cumulative percentages for the following grouped frequency distribution and write in the space provided.

Class	Midpoint	f	Cum f	Cum %
115–121		1		
108–114		3		
101–107		6		
94–100		10		
87–93		13		
80–86		16		
73–79		20		
66–72		29		
59–65		32		
52–58		24		
45–51		16		
38–44		12		
31–37		9	18	
24–30		5		
17–23		3		
10–16		1		
		$\sum f = N =$		

5. A census taken in a community to assess future educational needs revealed the following distribution of the number of children per family unit. Draw the appropriate graphical representation.

Number of children per family unit	f
10	1
9	0
8	5
7	12
6	22
5	103
4	184
3	297
2	431
1	158
0	220

6. Graph the three grouped frequency distributions constructed in Exercise 2.

7. It is frequently desirable to smooth the graphic representations of frequency distributions. One of the techniques most commonly employed is referred to as the method of moving averages. The procedures consist of the following steps:

a) Add two classes to each end of the frequency distribution. All four of these classes will have zero frequency associated with them.

b) Add a third column and label it "smoothed frequency." Enter a zero in the third column of the lowest class.

c) To find the smoothed frequency of the second class from the bottom, add together the frequencies in that class, the class above, and the class below, and divide by 3. Enter this value in the third column corresponding to the second from the lowest class.

d) Obtain the moving average for each class by first summing together the frequency within that class and the frequencies in the classes directly above and below that class and then by dividing by 3.

Shown below is the smoothed frequency distribution for data involving systolic blood pressures.

Class	f	Smoothed frequency
160–164	0	0.00
155–159	0	0.67
150–154	2	1.33
145–149	2	2.33
140–144	3	3.33
135–139	5	5.00
130–134	7	7.00
125–129	9	8.33
120–124	9	10.33
115–119	13	13.00
110–114	17	14.67
105–109	14	14.33
100–104	12	10.00
95–99	4	7.00
90–94	5	4.67
85–89	5	4.33
80–84	3	2.67
75–79	0	1.00
70–74	0	0.00

Construct a frequency curve for the smoothed grouped frequency distribution and superimpose upon it a frequency curve for the original grouped distribution. Describe the differences in the forms of the two distributions.

Answers:

1.

	Lowest and highest scores	Width of class	Apparent limits of the lowest class	Real limits of the lowest class	Midpoint of the lowest class
a)	9, 23	1	9	8.5–9.5	9
b)	15, 162	10	15–24	14.5–24.5	19.5
c)	81, 110	2	81–82	80.5–82.5	81.5
d)	–10, 52	5	–10 to –6	–10.5 to –5.5	–8
e)	0, 398	25	0–24	–0.5–24.5	12

2,3.

$i = 3$

Class	f	Cum f	Cum %
41–43	3	80	100
38–40	3	77	96
35–37	5	74	92
32–34	4	69	86
29–31	3	65	81
26–28	5	62	78
23–25	5	57	71
20–22	9	52	65
17–19	16	43	54
14–16	8	27	34
11–13	7	19	24
8–10	6	12	15
5–7	3	6	8
2–4	3	3	4

$i = 5$

Class	f	Cum f	Cum %
42–46	2	80	100
37–41	7	78	98
32–36	6	71	89
27–31	6	65	81
22–26	9	59	74
17–21	23	50	62
12–16	14	27	34
7–11	8	13	16
2–6	5	5	6

$i = 8$

Class	f	Cum f	Cum %
42–49	2	80	100
34–41	10	78	98
26–33	11	68	85
18–25	24	57	71
10–17	23	33	41
2–9	10	10	12

4.

Class	Midpoint	f	Cum f	Cum %
115–121		1	200	100.0
108–114		3	199	99.5
101–107		6	196	98.0
94–100		10	190	95.0
87–93		13	180	90.0
80–86		16	167	83.5
73–79		20	151	75.5
66–72		29	131	65.5
59–65		32	102	51.0
52–58		24	70	35.0
45–51		16	46	23.0
38–44		12	30	15.0
31–37		9	18	9.0
24–30		5	9	4.5
17–23		3	4	2.0
10–16		1	1	0.5

$$\sum f = N = 200$$

5.

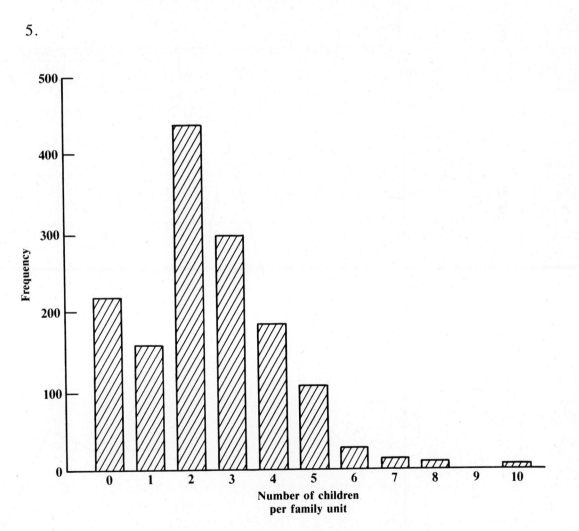

6. a)

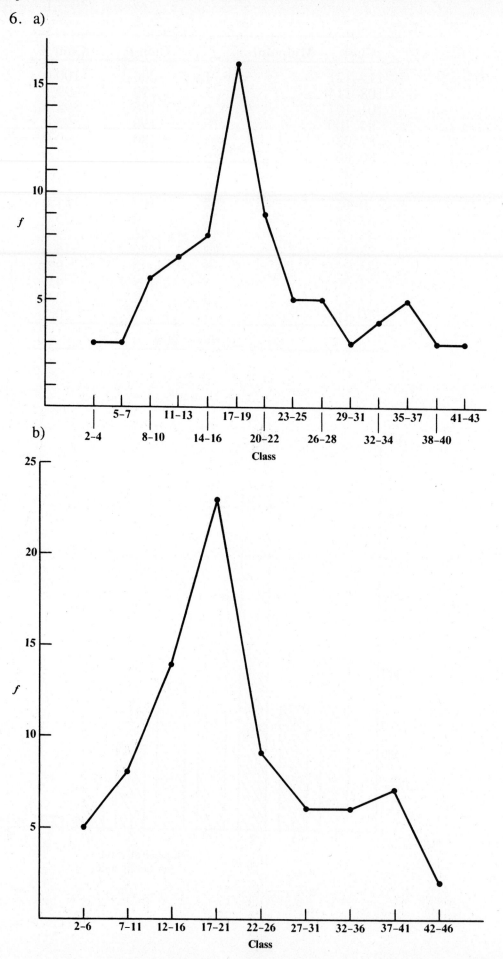

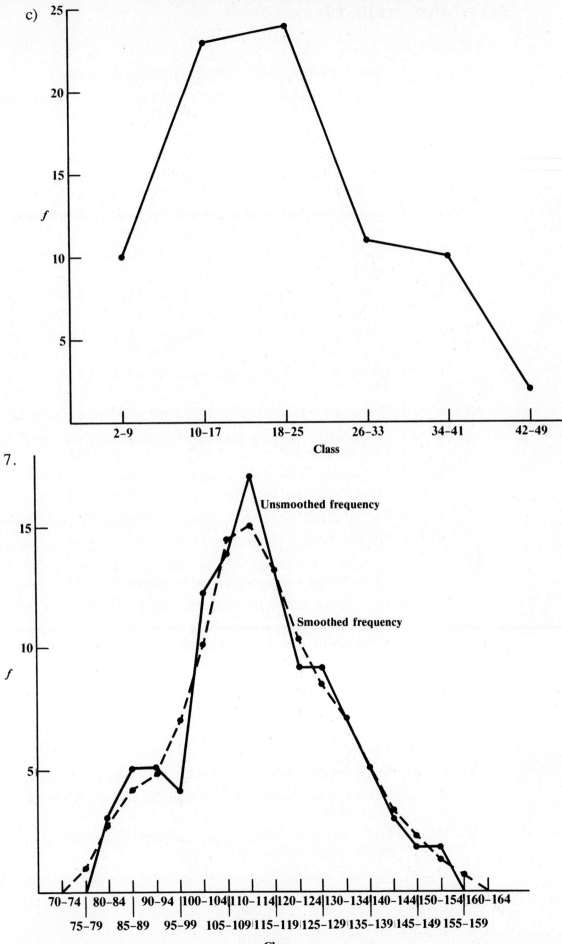

SELF-QUIZ: TRUE-FALSE

Circle T or F.

T   F   1. Grouping into classes involves collapsing the scale.

T   F   2. When we group scores into classes and calculate the frequency associated with each class, we construct a grouped frequency distribution.

T   F   3. A general rule to follow when grouping into classes is: Strive for as few classes as possible.

T   F   4. Once the decision concerning the number of classes is made, the procedure for assigning scores to classes is straightforward.

T   F   5. If there are 144 scores or potential scores in a distribution, $N = 96$, and the width of each class is 8, there are 12 classes.

T   F   6. An objective of grouping is to provide an economical and manageable organization of scores.

T   F   7. Referring to the cumulative percentage distribution in Selected Exercise 4 of this chapter, we see that 95% of cases are above a score of 100.5.

T   F   8. Referring to the cumulative frequency distribution of the same exercise, we see that 180 cases fall below a score of 90.

T   F   9. Referring to the same distribution, we see that 18 cases are above a score of 37.5.

T   F   10. Graphs are effective substitutes for statistical treatment of data.

T   F   11. A "gee whiz" chart is produced by eliminating the zero frequency from the vertical axis.

T   F   12. In graphing a frequency distribution, the height of the Y-axis should be between 0.70 and 0.80 the length of the X-axis.

T   F   13. The bar graph is employed with continuously distributed variables.

T   F   14. The bars in a bar graph should be touching rather than separated.

T   F   15. In graphic representations of frequency distributions, if the Y-axis is 200 mm in height, the X-axis should be approximately 150 mm.

T   F   16. The order of placing the categories along the ordinate in ordinally scaled data is arbitrary.

T   F   17. A frequency curve is employed with continuously distributed interval and ratio-scaled variables.

T   F   18. The bars in a histogram should be touching rather than separated.

T   F   19. Frequency is best represented by the height of the ordinate in frequency curves.

T   F   20. The normal curve is a symmetrical distribution.

T   F   21. Any symmetrical distribution is a normal curve.

T   F   22. It is not always possible to determine by inspection whether or not a distribution is skewed.

T   F   23. When groups of conditions are compared graphically, the experimental variable is represented along the Y-axis.

T   F   24. Random selection is essential in many studies if the results of the program are to be meaningful.

T   F   25. A table of scores with slash marks indicating frequency by each score is termed a grouped frequency distribution.

Answers: (1) T; (2) T; (3) F; (4) T; (5) F; (6) T; (7) F; (8) F; (9) F; (10) F; (11) T; (12) T; (13) F; (14) F; (15) F; (16) F; (17) T; (18) T; (19) F; (20) T; (21) F; (22) T; (23) F; (24) T; (25) F.

SELF-TEST: MULTIPLE-CHOICE

1. The real limits of the score 6 are:

   a) 6.0–7.0   b) 5.0–6.0   c) 5.5–6.5   d) 5.9–6.1   e) 5.95–6.05

2. When we arrange a set of scores in order of magnitude and indicate the frequency associated with each score, we have constructed:

   a) a grouped frequency distribution
   b) a frequency curve
   c) a histogram
   d) a bar graph
   e) an ungrouped frequency distribution

3. When we collapse a scale into mutually exclusive classes and indicate the frequency associated with each class, we have constructed:

   a) a cumulative frequency distribution
   b) a frequency distribution
   c) a grouped frequency distribution
   d) a histogram
   e) a nominal scale

4. It is generally agreed that most data in the behavioral sciences can be accommodated by how many classes?

   a) 5–10   b) 5–15   c) 10–15   d) 10–20   e) 15–30

5. If our lowest score were 19 and the highest score were 82 and we decided to group our scores into 8 classes, what would the width of each class be?

   a) 8   b) 9   c) 10   d) 19
   e) insufficient information to answer the question

6. The true limits of the class 80–84 are:

   a) 4   b) 5   c) 79–85   d) 80.5–83.5   e) 79.5–84.5

7. The midpoint of the class 20–23 is:

   a) 21   b) 21.5   c) 22   d) 22.5   e) none of the above

8. The midpoint of a class with a width of 5 is 20. The lower real limit is:

   a) 16.5   b) 17   c) 17.5   d) 18   e) 18.5

9. The midpoint of a class with a width of 7 is 12. The upper apparent limit is:

   a) 14   b) 14.5   c) 15   d) 15.5   e) 16

10. The apparent limits of a class are 15–32. The width of the class is:

    a) 32   b) 16   c) 15   d) 17   e) 18

11. If there are ten classes in a frequency distribution, and the apparent limits of the lowest class are 2–4, the apparent limits of the highest class are:

    a) 26–27   b) 26–28   c) 27–30   d) 27–29   e) 29–31

12. If we were to group a set of IQ scores into two mutually exclusive classes consisting of scores below 80 and those 80 and above:

    a) much of the information inherent in the original scores would be lost
    b) the classes would be too gross to preserve the original discrimination among scores
    c) there would be a gain in presentational economy
    d) all of the above
    e) none of the above

13. The midpoint of the class 12–21 is:

    a) 15   b) 15.5   c) 16   d) 16.5   e) 17

14. The width of the preceding class is:

    a) 12   b) 10   c) 9   d) 10.5   e) 9.5

Basing your answer on the following distribution, answer Questions 15 through 23.

Class	f	Cum f	Cum %
35-39	2	50	100
30–34	3	48	96
25–29	7	45	90
20–24	9	38	76
15–19	14	29	58
10–14	8	15	30
5–9	5	7	14
0–4	2	2	4
	$N = 50$		

15. The frequency of 3 in the class 30–34 means:

    a) 3 frequencies are at the upper real limit of the class
    b) 3 frequencies are at the lower real limit of the class
    c) 3 frequencies are uniformly spread out throughout the class
    d) 3 frequencies are at the lower apparent limit of the class
    e) 3 frequencies are at the upper apparent limit of the class

16. The cum f of 48 means that 48 cases:

    a) fall below a score of 35
    b) fall below a score of 30
    c) fall below a score of 29.5
    d) fall below the midpoint of the class
    e) fall below a score of 34.5

17. The cum % of 76 means that 76% of the cases:

    a) fall below a score of 24
    b) fall below a score of 20
    c) fall below a score of 24.5
    d) fall below a score of 19.5
    e) fall below a score of 22

18. The score corresponding to a cumulative percentage of 58 is:

    a) 19.5   b) 17   c) 19   d) 15   e) 14.5

19. The score corresponding to a cum f of 7 is:

    a) 10   b) 9.5   c) 7   d) 15   e) 5

20. The score *above* which 70% of the cases fall is:

    a) 14.5   b) 14   c) 12   d) 10   e) 9.5

21. A score of 18 has:

    a) between 16% and 30% of the cases below it
    b) between 58% and 76% of the cases below it
    c) between 14% and 30% of the cases below it
    d) between 30% and 58% of the cases below it
    e) insufficient information to answer the question

22. The score at the lower real limit of the lowest class is:

    a) 2   b) 4   c) 0   d) −.5   e) .5

23. The score at the midpoint of the highest class is:

    a) 36   b) 36.5   c) 37   d) 37.5   e) 39

24. In a grouped frequency distribution, all the scores in a class are represented by:

    a) the upper real limit   b) the lower real limit   c) the midpoint
    d) the upper apparent limit   e) the lower apparent limit

25. Under which of the following conditions are the midpoint and the apparent limits of a class identical?

    a) when the frequency distribution is bilaterally symmetrical
    b) when measurements are continuous and the range of the distribution is less than 10.0
    c) when data are discrete and the distribution contains less than 20 different values
    d) when the midpoints are multiples of the width of the class
    e) when the width of the class is equal to 1

26. A student prepares a frequency distribution with six categories. Which of the following problems is most likely to arise?

    a) Extreme values will be given undue weight in computing central tendency.
    b) Final statistical values will contain excessive error because of grouping.
    c) Final statistical values will contain excessive error because of rounding.
    d) Hand computation will be unwieldly and time consuming.
    e) Scores that do not fall in the class will have to be eliminated.

27. In plotting which of the following curves do we use the upper real limit of the class rather than the midpoint?

    a) frequency curve   b) histogram   c) cumulative frequency curve
    d) none of the above   e) all of the above

28. Another name for the X-axis is:

    a) vertical axis   b) ordinate   c) abscissa   d) dependent variable
    e) frequency scale

29. In a cumulative frequency curve, each frequency is plotted over:

    a) the midpoint of the class
    b) the upper real limit of the class
    c) the lower real limit of the class
    d) the apparent limits of the class
    e) the frequency corresponding to the class

30. In comparing frequency distributions, which type of graph is preferable?

    a) bar graph   b) histogram   c) frequency curve
    d) cumulative frequency curve   e) ogive

31. Most generally a graph may be considered:

    a) a substitute for statistical treatment of data
    b) a visual aid for thinking about data
    c) a dependable means of avoiding misinterpretations of data
    d) a last resort of the uninformed
    e) a pictorial technique that almost always leads to confusion

32. A graph may be employed to mislead the reader by:

    a) manipulation of the vertical axis
    b) manipulation of the horizontal axis
    c) making the initial entry on the ordinate some value other than zero
    d) all of the above
    e) none of the above

33. With nominally or ordinally scaled data it is common to employ:

    a) histograms   b) frequency curves   c) ogives   d) normal curves
    e) bar graphs

34. Employing the usually accepted graphing convention, one finds that if the ordinate is 90 units, the abscissa is:

    a) 120 units   b) 180 units   c) 45 units   d) 67.5 units   e) 75 units

35. A frequency curve characterized by a piling up of scores in the center of the distribution is called:

    a) platykurtic   b) mesokurtic   c) leptokurtic   d) asymmetrical
    e) centrokurtic

36. A cumulative frequency distribution of normally distributed data will yield:

    a) a J-curve   b) a histogram   c) a symmetrical distribution
    d) an ogive   e) none of the above

37. In a bar graph:

    a) the bars should be touching one another
    b) no graphing conventions need be employed
    c) we may convert to a frequency curve by joining the midpoints of the bars by straight lines
    d) the bars should be separated
    e) none of the above

38. If in a frequency curve there are relatively few frequencies at the right end of the horizontal axis when compared with the left end, the curve is:

    a) positively skewed   b) an ogive   c) symmetrical   d) platykurtic
    e) negatively skewed

39. In graphs comparing various groups on a criterion variable, the criterion variable

    a) is generally represented on the X-axis
    b) depends on whether or not the criterion variable is quantified
    c) is generally represented on the Y-axis
    d) depends on the purpose of the investigator
    e) none of these

40. The difference between a bar graph and a histogram is that:

    a) a bar graph is less precise
    b) a bar graph is readily converted into a frequency curve
    c) the bars on a bar graph are contiguous
    d) the histogram is employed with interval- and ratio-scaled variables
    e) none of the above

41. If we arbitrarily set the width of each bar at one unit in the histogram and bar graph, the total area of all bars is equal to:

    a) N
    b) $\sum f/N$
    c) the area of the class with the greatest frequency multiplied by i, the width of the class
    d) N/i
    e) Ni

42. In graphic representations of frequency distributions, it is generally desirable to represent the frequency of a given category in terms of the:

    a) length of the abscissa
    b) height of the ordinate
    c) area between selected points on the abscissa
    d) number of classes employed
    e) height of the ordinate for the class containing the highest associated frequency

Basing your answers on the following information, answer Questions 43 through 46. A frequency curve is constructed in which there are ten classes, $i = 5$. The abscissa or X-axis is laid out along a line 160 mm in length.

43. The height of the ordinate is approximately:

    a) $26^2/_3$ mm   b) 32 mm   c) 7 mm   d) 120 mm   e) 160 mm

44. The number of millimeters per class is:

    a) 32   b) $26^2/_3$ c) 7   d) 12   e) 16

45. The number of millimeters per frequency is:

    a) 7   b) 5   c) 6   d) 9   e) insufficient information

46. In a cumulative frequency curve, the height of the ordinate is approximately:

    a) 32 mm   b) 7 mm   c) 600 mm   d) 160 mm   e) 120 mm

Employing the graph shown here and assuming that each score is represented by one unit on the scale of frequency and one equal unit on the scale of scores, answer Questions 47 through 50.

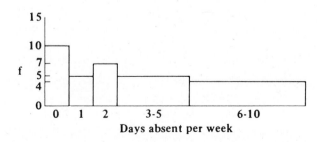

Days absent per week

47. The frequency associated with 0 days absent from work is:

    a) 0   b) 5   c) 10   d) 5
    e) none of the above

48. The frequency associated with 3–5 days absent from work is:

    a) 1.67   b) 5   c) 10   d) 15   e) 20

49. The frequency associated with 3–10 days absent from work is:

    a) 2.9   b) 9   c) 31   d) 35
    e) cannot determine without seeing actual frequency distribution

50. N is equal to:

    a) 24.9   b) 31   c) 54   d) 57
    e) cannot determine without seeing actual frequency distribution

Answers: (1) c; (2) e; (3) c; (4) d; (5) a; (6) e; (7) b; (8) c; (9) c; (10) e;
(11) e; (12) d; (13) d; (14) b; (15) c; (16) e; (17) c; (18) a; (19) b; (20) a;
(21) d; (22) d; (23) c; (24) c; (25) e; (26) b; (27) c; (28) c; (29) b; (30) c;
(31) b; (32) d; (33) e; (34) a; (35) c; (36) d; (37) d; (38) a; (39) c; (40) d;
(41) a; (42) c; (43) d; (44) e; (45) e; (46) e; (47) c; (48) d; (49) d; (50) d.

4

Percentiles

BEHAVIORAL OBJECTIVES

Conceptual Objectives

1. Define a percentile rank and state its function.

2. Find the percentile rank of a score from a cumulative percentage graph.

3. Use a cumulative percentage graph to find the score corresponding to a given percentile rank.

4. Using the formula in the text, calculate percentile ranks directly from scores. Explain this method of calculating percentile ranks.

5. Interpret the percentile rank of a score.

6. Calculate the score corresponding to a particular percentile rank.

7. Explain the function of reference groups in relation to percentile ranks.

Procedural Objectives

1. With the aid of interpolation techniques, calculate cumulative frequencies of scores occurring within each class of a frequency distribution.

2. Determine the percentile ranks of scores occurring within classes.

3. Calculate the cumulative frequency that is associated with a specific percentile rank. Derive actual scores from the cumulative frequencies.

CHAPTER REVIEW

A parachuting school listed Alan as scoring at the 95th percentile on his first jump. What does this statement signify? In a word, it means that Alan scored higher than 95% of those in the comparison group. We can define the *percentile rank* of a score as the percentage of cases in a comparison group whose scores fell below the score (Alan's) being considered.

From their records of first jumps, the instructors at the parachuting school had assigned a percentile rank to all possible scores. When Alan scored 68 on his first attempt, the instructor consulted her chart to find that the corresponding percentile rank was 95.

Suppose, however, that the instructor did not have a table of all possible scores with their corresponding percentile ranks. Rather, she had a grouped frequency distribution of scores made by all students who preceded the present class. To calculate the percentile rank corresponding to Alan's score, the instructor would have to construct a cumulative frequency distribution, find the class that contained Alan's score, and interpolate within that class.

Let us suppose that the following information is available:

1. Alan's score falls in the class with the apparent limits of 66 to 70 and the real limits of 65.5 to 70.5.

2. The cumulative frequency to the lower limit of Alan's class is 186.

3. The total N in the distribution is 200.

4. Eight cases or frequencies lie within this class that contains 5 different scores: 66, 67, 68, 69, and 70.

   With this information, we are ready to calculate the percentile rank of Alan's score of 68.

Step 1: Alan's score of 68 lies 2.5 units above the lower real limit of the class, that is $(68 - 65.5)/5 = 2.5$.

Step 2: Since there are 5 score units within the entire class $(70.5 - 65.5 = 5)$, his score is 2.5/5 the way through the class.

Step 3: Since there are 8 cases or frequencies within the entire class, his score is $2.5/5 \times 8 = 4$ frequencies above the real limit of the class.

Step 4: With 186 cumulative frequencies below the real limit of Alan's class and 4 additional frequencies up to his score, the cumulative frequency of his score is $186 + 4 = 190$.

Step 5: His percentile rank, then, is $(186 + 4)/200 \times 100 = 95$. Using Formula (4.1) in the text, we obtain the same result:

$$\text{Percentile rank of 68} = \left(\frac{186 + [(68 - 65.5)/5] \times 8}{200} \right) \times 100$$
$$= 190/200 \times 100 = 95$$

As we were working through the preceding steps, we made a major assumption: The cases within a particular class are evenly distributed throughout that class. If the cases had not been evenly distributed, 68 might not have been the fourth case within the class. For example, what if all 8 cases within the class were scores of 70? In this unlikely event, our procedures would have been somewhat in error. A loss of some precision is the price we must pay to achieve the presentational economy introduced by grouping scores into classes. By and large, the assumption of an even spread throughout the class will not seriously affect the integrity of our calculations.

We have just seen how to calculate the percentile rank of a score. In this case, the score is known but the percentile rank is not. But what if we are given the percentile rank and wish to learn the corresponding score? Now our solution involves the percentile rank as the known value and the score as the unknown.

To find the score corresponding to a given percentile, we must first multiply the sample size by the percentile rank. This provides the cumulative frequency corresponding to that percentile rank. Next, looking in the cumulative frequency column, we find the class that contains the desired cumulative frequency. The desired score will be located somewhere in the class between the cumulative frequency at the lower real limit of the class and the cumulative frequency at the upper real limit of that class. Interpolation is then used to locate precisely the score within that class.

For simplicity and as a check of our prior computations, let us use the values with which we are already familiar. In the parachuting example, what is the score at the 95th percentile? To answer the question, we use the following formula:

$$\text{Score at a given percentile} = X_{ll} + i(\text{cum} f - \text{cum} f_{ll})/f_i$$

Recall that X_{ll} = the score at the lower real limit of the class containing cum f; i = the width of the class, cum f = the cumulative frequency of the score, cum f_{ll} = the cumulative frequency at the lower real limit of the class containing cum f, and f_i = the number of cases within the class containing cum f.

To find the score corresponding to a known percentile rank:

Step 1: Multiply the sample size (200) by the decimal form of the percentile rank (.95) to obtain the cumulative frequency corresponding to that rank—$200 \times .95 = 190$.

Step 2: Find the score (X_{ll}) of the lower real limit of the class containing that cumulative frequency. In this example, it is 65.5.

Step 3: Find the cumulative frequency (f_{ll}) at the lower real limit of that class. It is 186.

Step 4: By inspection, determine the number of cases or frequencies within that class. As you can see, it is 8.

Substituting the appropriate values into the formula, we obtain

$$\text{Score at the 95th percentile} = 65.5 + 5(190 - 186)/8 = 65.5 + 2.5 = 68$$

Knowing the reference group is essential for interpreting the meaning of a percentile rank. Suppose you were told that a 28,500-foot deep oil well drilled in Pecos County, West Texas, in 1972 is at the 99th percentile of oil well depths. How impressed are you with this statement? This depends on what wells were in the reference group. If the reference group was restricted to other oil wells in Pecos County, the statement would not be nearly as impressive as the actual facts of the matter. The reference group consisted of all the deepest wells in the world. That's like being in the 99th percentile when the reference group is all the people with IQs in the genius category!

Estimating, by Inspection, the Percentile Ranks of Scores and the Scores Corresponding to Percentiles

As you move further into this course, you are almost certain to experience, on occasion, the frustration of obtaining an answer to a problem that is far wide of the mark—a misplaced decimal, the wrong value entered into the calculator, an incorrect key pressed, and so forth, *ad infinitum ad nauseum.* In order to recognize when you have made a big error, you should make it a habit to estimate the target statistic before doing any of the calculations. There are simple observational procedures by which most statistics can be estimated, often with surprising accuracy. Procedures will be presented throughout this Study Guide for guesstimating statistics. When doing course assignments or statistical analyses on your own, you should check your calculated answers against your guesstimates. If they are far apart, chances are you made a computational error.

Table 4.1 shows a grouped and cumulative frequency distribution ($i = 5$) of 110 scores. Let's suppose you wish to guesstimate the score at the 40th percentile. You should engage in the following conversation with yourself. "First I multiply, in my head, .4 (the percentile) times 110 (the N). That's about 44. I look at the cumulative frequency column and see that 44 is in the class with the lower real limit of 109.5. However, it's barely into the class (one above the highest frequency in the preceding class). That will put the score at the 40th percentile pretty close to the lower real limit of the class. In other words, it's somewhere around and sightly more than 109.5."

Table 4.1   Grouped Frequency Distribution and Cumulative Frequency Distribution of IQ Scores Obtained by 110 School-Age Children

Class	f	Cumulative f
150–154	2	110
145–149	2	108
140–144	3	106
135–139	5	103
130–134	7	98
125–129	9	91
120–124	9	82
115–119	13	73
110–114	17	60
105–109	14	43
100–104	12	29
95–99	4	17
90–94	5	13
85–89	5	8
80–84	3	4

$$N = 110$$

Armed with the knowledge that the score at the 40th percentile is somewhere in the neighborhood of 109.5, you go ahead with your calculations. Don't be surprised to learn that your guesstimate was pretty close to the correct answer—109.79 to be exact.

Now let's try the obverse guesstimation. You wish to know the percentile rank of a score of 127. Your soiloquy should go something like this: "In what class does a score of 127 lie? Ah, it's in the class with the real limits of 124.5 and 129.5. The cumulative frequency to the lower limit of the class is 82. But 127 appears to be about halfway through the class. So its corresponding cumulative frequency is $82 + 9/2 = 86.5$. When this is divided by 110 and multiplied by 100 on my trusty calculator (or divided by 1.10 to avoid the multiplication by 100), I get 78.64. Now what's the correct answer?

When you go through the calculations, you may be shaken to learn that your guesstimate was precisely on the mark; a bit of good luck due to the fact that 127 was exactly at the midpoint of the class. This will not happen often, but one thing I can guarantee—if you practice hard at guesstimating statistics before calculating them, you will almost always spot a grossly incorrect calculation!

SELECTED EXERCISES

1. For the frequency distributions constructed in Selected Exercise 2, Chapter 3 of the Study Guide, find the scores at the following percentile ranks: 16, 25, 86, 98. (See p. 39.)

2. For the frequency distributions constructed in Selected Exercise 2, Chapter 3 of the Study Guide, find the percentile ranks corresponding to the following scores: 5, 18, 31, 44.

3. Using the grouped frequency distribution shown in Selected Exercise 7, Chapter 3 of the Study Guide, find the percentile ranks of the following scores: 154, 115, 70.

4. Using the grouped frequency distribution shown in Selected Exercise 7, Chapter 3 of the Study Guide, find the scores corresponding to the following percentile ranks: 1, 16, 58, 95.

Answers:

1.

Percentile	Score			
	$i = 1$	$i = 3$	$i = 5$	$i = 8$
16	10.8	10.84	11.38	10.47
25	14.5	13.88	14.00	12.98
86	34.3	34.35	34.67	34.14
98	41.9	41.9	42.5	43.1

2.

Score	Percentile rank			
	$i = 1$	$i = 3$	$i = 5$	$i = 8$
5	5.00	4.38	4.38	5.47
18	42.50	43.75	42.38	43.12
31	81.25	80.62	80.50	80.70
44	100.00	100.00	98.75	98.28

3.

Score	Percentile rank
154	99.82
115	55.73
70	0

4.

Percentile rank	Score
1	81.33
16	99.75
58	115.96
95	142.00

SELF-QUIZ: TRUE-FALSE

Circle T or F.

T F   1. A score by itself is meaningless.

T F   2. John G., who had a percentile rank of 3 on a standardized test, scored higher than Mike F., who had a percentile rank of 90.

T F   3. In a cumulative percentage graph, the cumulative percentage is usually shown along the horizontal axis.

T F   4. The cumulative percentage distribution is useful in obtaining a percentile rank directly.

T F   5. In interpolating within a class, we employ the lower apparent limits of the class.

T F   6. If a score lies in the class 124.5–129.5, we must know the cumulative frequency at the upper limit of the class to determine the percentile rank of the score.

T F   7. A score of 110 has an associated cumulative frequency of 90, with $N = 120$. The percentile rank of the score is 91.67.

T F   8. If $N = 200$, the cumulative frequency at the 80th percentile is 160.

T F   9. In finding the score corresponding to a given percentile rank, you must interpolate from the cumulative frequency scale to the scale of scores.

T F   10. A score of 85 has an associated cumulative frequency of 200 in which $N = 400$. The percentile rank of the score is 50.

T F   11. Just as a score is meaningless in the abstract, so also is a percentile rank.

T F   12. Many standardized tests used in education and industry publish separate norms for various reference groups.

T F   13. Referring to Table 4.2 in the text, we find that a person would have to score higher in social work to achieve the 65th percentile than an individual in the social sciences.

T F   14. When calculating percentile ranks for grouped data, we make use of the apparent limits of a class.

T F   15. So far as inflation is concerned, Italy is at the highest 10 percent of all nations. Italy's percentile rank is 10.

T F   16. Interpolation is usually necessary to determine a percentile rank of a score from a grouped frequency distribution.

T F   17. To find a score from a cumulative frequency distribution, interpolation is *not* a required procedure.

T F   18. The number of dead from skirmishes in the Mideast on November 11 was assigned a percentile rank of 8. No further information is necessary in order to understand the full meaning of this statement.

Answers: (1) T; (2) F; (3) F; (4) T; (5) F; (6) F; (7) F; (8) T; (9) T; (10) T; (11) T; (12) T; (13) F; (14) F; (15) F; (16) T; (17) F; (18) F.

SELF-TEST: MULTIPLE CHOICE

1. If the percentile rank of a score of 20 is 65, we may say that:

   a) 20% of a comparison group scores at or below 65
   b) 20% of a comparison group scored above 65
   c) 65% of a comparison group scored above 20
   d) 65% of a comparison group scored at or below 20
   e) none of the above

2. When determining the score corresponding to a given percentile rank by interpolating within the class, an important assumption is:

   a) that the cases or frequencies are normally distributed throughout the class
   b) that the cases or frequencies are evenly distributed throughout that class
   c) that the grouped frequency distribution of scores is rectangular
   d) all of the above
   e) none of the above

3. Assume that there are ten cases in the class with the apparent limits of 1–5. The score that corresponds to a cumulative frequency of 7 is:

   a) 4.5   b) 2.67   c) 3.0   d) 3.5   e) 4.0

4. In the preceding example, the cumulative frequency of a score of 2 is:

   a) 1   b) 3   c) 4   d) 5   e) 8

Basing your answers on the cumulative percentage distribution shown here, answer Multiple-Choice Problems 5 through 8.

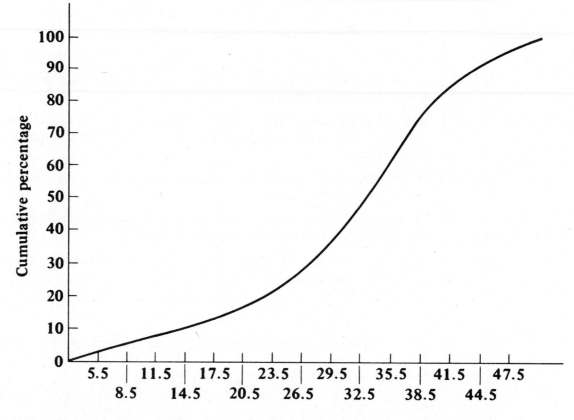

(Drawing for items 5–8)   Score at the upper real limit of each class

5. The percentile rank of a score of 30 is approximately:

   a) 38   b) 42   c) 28   d) 30   e) 24

6. The percentile rank of a score of 43 is approximately:

   a) 30   b) 88   c) 84   d) 95   e) 33

7. The score corresponding to a percentile rank of 60 is:

   a) 30.5   b) 34   c) 35.5   d) 37   e) 38.5

8. The score corresponding to a percentile rank of 10 is:

   a) 47.5   b) 7   c) 14.5   d) 8.0   e) none of the above

9. When we convert scores to percentile ranks and vice versa, we are performing:

   a) a curvilinear transformation   b) a cumulative transformation   c) a linear transformation   d) a grouping conversion   e) a reversal conversion

Basing your answers on the following grouped frequency and cumulative distributions of scores, answer Multiple-Choice Problems 10 through 15.

	f	Cum f
36–41	3	50
30–35	5	47
24–29	8	42
18–23	16	34
12–17	10	18
6–11	6	8
0–5	2	2

$$i = 6, N = 50$$

10. The score at the 40th percentile is in the class with the apparent limits of

   a) 6–11   b) 12–17   c) 18–23   d) 24–29   e) 30–35

11. The score with a percentile rank of 70 is in the class with the apparent limits of:

   a) 36–11   b) 30–35   c) 24–29   d) 18–23   e) 6–11

12. The score with a percentile rank of 4 is:

   a) 2   b) 3.5   c) 5   d) 5.5   e) 6

13. The score with a percentile rank of 50 is approximately:

   a) 41.5   b) 22.7   c) 20.1   d) 23.2   e) 18.6

14. A score of 13 has an approximate percentile rank of:

   a) 17.0   b) 25.0   c) 15.3   d) 21.0   e) 17.2

15. A score of 23.5 has a percentile rank of:

   a) 32.0   b) 16.0   c) 68.0   d) 65.4   e) 84

16. An individual's score has a percentile rank of 90 on an achievement exam. This means that he or she:

   a) has twice as many items right as someone with a percentile rank of 45
   b) performed as well as or better than 90% of the group taking the test
   c) placed ninetieth in the group taking the test
   d) answered 90% of the total number of items correctly
   e) impossible to say without further information

17. In a sample in which $N = 120$, John S. obtained a score of 30, placing him at the 90th percentile. Sally Q. obtained a percentile rank of 45. Her score is:

   a) 54   b) 15   c) 60   d) correct answer not given
   e) impossible to say without further information

18. A percentile rank of a score is defined symbolically as:

   a) $\dfrac{\text{cum } f}{N} \times 100$ b) $\dfrac{\text{cum } f \cdot N}{100}$ c) $\dfrac{\text{cum } f}{N \cdot 100}$

   d) $\dfrac{\text{cum } fN}{100}$ e) $\dfrac{\text{cum } f}{100} \times 100$

19. Mark G. obtains a score of 98 on a standard scholastic test. You should:

    a) commend him for obtaining such a high score
    b) ask him why he failed to score higher
    c) assume that he obtained 98 correct answers out of 100
    d) all of the above
    e) none of the above

20. Fred S. obtained a score of 80 on a statistics test, placing him at the 88th percentile. If ten points were added to each score in the distribution, his new score of 90 would be at the:

    a) 90th percentile
    b) 98th percentile
    c) 88th percentile
    d) none of the above
    e) impossible to answer without additional information

21. Which of the following statements is most likely to be correct for a given distribution?

    a) A percentile rank of 60 is twice as great as a percentile rank of 30 when transformed into raw score values.
    b) The range in raw score values between percentile ranks of 20 and 30 equals that between percentile ranks of 40 and 50.
    c) The range in raw score values up to the 50th percentile equals that above the 50th percentile.
    d) There are four times as many cases above the 40th percentile as above the 10th percentile.
    e) None of the above.

Answers: (1) d; (2) b; (3) e; (4) b; (5) a; (6) b; (7) c; (8) c; (9) c; (10) c; (11) c; (12) d; (13) c; (14) d; (15) c; (16) b; (17) e; (18) a; (19) e; (20) c; (21) e.

5

Measures of Central Tendency

BEHAVIORAL OBJECTIVES

1. Determine the reason for replacing the term "average" with "measures of central tendency." Define each measure of central tendency.

2. Specify the two characteristics of frequency distributions for which statisticians have developed quantitative descriptive methods.

3. Define the arithmetic mean in words as well as in algebraic form. Calculate the arithmetic mean for data sets as well as ungrouped and grouped frequency distributions.

4. List the three properties of the mean. Identify an array of scores.

5. Distinguish between a mean and a weighted mean in words and in algebraic form.

6. Describe the median algebraically and with words. Calculate the median from an odd and even number of schools in an array. Explain the characteristics of the median.

7. Guesstimate the mean and the median by inspecting a frequency distribution.

8. Ascertain and define the mode.

9. Compare and contrast the mean, median, and mode in terms of their various characteristics. Explain how skewness affects the median and the mean.

CHAPTER REVIEW

The title of this chapter, "Measures of Central Tendency," may suggest that you are entering a rather hostile alien region. However, if the word "average" were substituted for "central tendency," you might realize that you were entering territory that you had explored on numerous occasions.

Central tendency is indeed another expression for average. Statisticians have come to prefer the term central tendency because of the wide use and abuse of the term average. When you open a magazine to the latest diet fad, you may read than an 18-year-old woman with a weight of 119 pounds and a height of 65 inches has an average daily need of 2100 calories. You will realize that the person responsible for these figures was citing a measure of central tendency. But which one? The mean, median, or mode?

The fact of the matter is that there is more than one way to measure central tendency. It is for this reason that confusion over the usage of this word occurs in all walks of life. Indeed, some people find this confusion to their liking because they can take advantage of it to pull off seemingly magical numerical "sleights of mind." The differences between the mean and median are often central to statistical legerdemain, also known as lying with statistics.

The first measure of central tendency, and the easiest to ascertain, is the *mode*. By definition, it is the score (or midpoint of a class) that occurs most frequently in a distribution. The method of finding the mode in any given distribution involves nothing more than inspection.* If you are dealing with a frequency or grouped frequency distribution, you only need to scan the frequency column for the largest value. The score or midpoint of the class that is associated with

---

*There is also a mathematical mode that is obtained by computation. However, it is so rarely used that we shall not concern ourselves with it here.

that frequency is the mode. If you are given a list of unorganized scores, you can either construct your own frequency distribution or examine the list for the score that is repeated most often.

A second measure of central tendency is the *median*. Since you are already acquainted with the procedures for finding the score at a given percentile, you should be able to calculate the median without difficulty. Why? The median is that score or potential score in a distribution above and below which one-half of the frequencies fall. In other words, it is the middle score—the score at the 50th percentile.

In finding the median, we use a modified version of the formula for calculating the score at a given percentile, the percentile in this case being 50. Thus

$$\text{Median} = X_{ll} + i(N/2 - \text{cum } f_{ll})/f_i$$

where X_{ll} = the score at the lower real limit of the class containing the cumulative frequency of $N/2$, i = the width of the class, N = the total number of cases in the distribution, cum f_{ll} = the cumulative frequency at the lower real limit of the class containing the cumulative frequency of $N/2$, and f_1 = the number of cases within the class containing $N/2$.

Table 5.1 shows the grouped frequency distribution and cumulative frequency distributions of the data in Table 3.3 of this manual. This table combines the percentages of chromosomal instabilities among 21 patients with thyroid cancer and six control subjects (Hsu et al., 1981). The calculation of the median percentage of chromosomal instabilities follows.

Step 1: Multiply the sample size (27) by the decimal form of the 50th percentile (0.50) to obtain the cumulative frequency $N/2$ corresponding to the median—27/2 or $27 \times .50 = 13.5$.

Step 2: Find the score X_{ll} of the lower real limit of the class containing that cumulative frequency. In this example, it is 5.35.

Step 3: Find the cumulative frequency f_{ll} at the lower real limit of that class. It is 11.

Step 4: By inspection, determine the number of cases or frequencies f_i within that class. As you can see, it is 3.

Step 5: Find the width i of the class by subtracting the score at the lower real limit of any class from the upper real limit of that class—$i = 3.15 - .95 = 2.2$. This applies only when all classes have the same width. Otherwise, find the width of the class that contains the score at the median.

Step 6: Substitute the values obtained in the above steps in the formula for calculating the median of a grouped frequency distribution:

$$
\begin{aligned}
\text{Median} &= X_{ll} + i(N/2 - \text{cum } f_{ll})/f_i \\
&= 5.35 + 2.2(13.5 - 11)/3 \\
&= 5.35 + 1.83 \\
&= 7.18
\end{aligned}
$$

Table 5.1   Grouped Frequency and Cumulative Frequency Distributions of Percentages of Chromosomal Instabilities Among 21 Patients with Thyroid Cancer and 6 Control Subjects

Real limits of class	f	Cum f	Cum %
16.35–18.55	2	27	100.00
14.15–16.35	0	25	92.59
11.95–14.15	8	25	92.59
9.75–11.95	2	17	62.96
7.55–9.75	1	15	55.56
5.35–7.55	3	14	51.85
3.15–5.35	5	11	40.74
0.95–3.15	6	6	22.22

$$N = 27$$

$$\text{Median} = 5.35 + 2.2(13.5 - 11)/3 = 5.35 + 1.83 = 7.18$$

Source: Hsu et al., 1981.

The final measure of central tendency, the arithmetic mean, is probably the most familiar. This statistic is often used in sports (batting averages, average yards kicked by a punter, average number of goals scored against an ice hockey goalie, average number of points scored by a basketball forward, and so forth), and it is the method by which you have been averaging your grades for so many years. Briefly stated, the arithmetic mean is the sum of the scores or values of a variable divided by the number of cases. For example, every time a baseball player gets a base hit (single, double, triple, or home run), he or she is awarded one point. To find the arithmetic mean, you divide the number of points earned by the number of times officially at bat. Recall, a walk does not constitute an official time at bat.

Algebraically stated, the definition becomes

$$\bar{X} = \frac{X_1 + X_2 + \cdots + X_N}{N} = \sum X/N$$

where $\bar{X}$ = the mean (X bar), N = the number of observations or cases, and $\sum$ is the mathematical verb directing us to sum all of the measurements.

To practice using the formula for calculating the mean, substitute into the formula the appropriate numbers from the following example. Recall the study by Hsu et al. (1981) in which the percentage of cells showing chromosome instability was obtained from 21 patients with thyroid cancer and 6 control subjects. The scores or percentages among the control subjects were 4, 1, 5, 7, 3, 3. The sum of the values equals 23. Since the number of scores (N) equals 6, the mean is 23/6 = 3.83, rounded to the second decimal place. What about the 21 patients with thyroid cancer? Their results are shown in the following table:

13.9	12.0	17.1	10.9	13.0
13.4	9.3	11.3	6.0	5.2
12.2	14.0	14.0	5.0	18.0
2.0	2.0	13.6	3.2	2.0
5.4				

The sum of the values of the variable for this data set equals 203.5. Since N = 21, the mean is 203.5/21 = 9.69. As you can see, for this sample, the mean of the patients is considerably higher than that of the control subjects. However, without using inferential statistics, we do not know if the difference between groups is sufficiently large to conclude that the samples were drawn from populations with different means. We'll reexamine these data from this perspective in Chapter 13.

As you have seen, it is often desirable to construct a frequency distribution to eliminate adding repeatedly the same scores that often occur when large amounts of data are involved. To simplify calculations, you should use the following formula for finding the mean when the data are arranged in the form of ungrouped and grouped frequency distributions:

$$\bar{X} = \Sigma fX/N$$

where X is the midpoint of the class for grouped frequency distributions.

If the frequency of a score (or midpoint of a class) of 10 is 4, you multiply 4 times 10 rather than add 10 four times. Suppose your frequency is 11 and the midpoint of a class is 25. You would multiply 11 times 25, add that number to the sum of your other midpoints times their frequencies, and divide by the sum of all the frequencies, which is N.

Finding the Weighted Mean

At times when a number of different samples are taken, you calculate the mean of each, and then you need to know the combined, overall, or grand mean of all the samples. If each is based on the same sample size, the calculations are easy; merely add the various means together to find their sum and divide by the number of means. For example, if you purchased 60 shares each of four different common stocks and paid the following means per share—$8.00, $15.50, $19.00, and $22.00—the overall mean is $(8 + 15.5 + 19 + 22)/4 = \$14.125$.

However, if the sample size (number of shares purchased) is not the same for each transaction, you must use a procedure that takes into account the different N's. In a situation like this, you must obtain what is called a weighted mean. The formula for the weighted mean is

$$\bar{X} = \Sigma f\bar{X}/N$$

where $\Sigma f = N$ = the total number of means over all groups and $\bar{X}$ = (the mean of each group of scores).

To illustrate, the following table shows our purchase of common stock in one company at four different times and at four different prices. What is the overall mean price that we paid?

$\bar{X}$, mean price per share	f, number of shares purchased	$f\bar{X}$, product of f times $\bar{X}$
16.5	80	1320
24.0	120	2880
28.0	130	3640
40.5	200	8100
	$\Sigma f = N = 530$	$\Sigma f\bar{X} = 15{,}940$

The weighted mean, $\bar{X}_W$, equals $15{,}940/530 = 30.08$ (rounded).

Now that you are versed in the calculation of the median and mean, let's compare the two in terms of their properties and their application to the analysis of data.

To begin with, every score participates in the determination of the mean. This property permits the direct calculation of the sum of all the scores when we know only the mean and the sample size. Thus, if we are told that the 135 employees of XYZ corporation earn a mean weekly income of $600, we know that the total weekly payroll is $600 \times 135 = \$81,000$ ($\bar{X} \cdot N = \sum X$). Similarly, if you know that Ty Cobb batted .3667 (mean) in 11,429 official at bats (N), you can ascertain the number of hits he made ($\sum X$): $\sum X = .3667 \cdot 11,429 = 4,191$. That enviable number of career hits has since been surpassed by Pete Rose.

This very same characteristic—the fact that all scores participate in the mean—makes the mean very sensitive to extreme measurements when these measurements are not balanced on both sides of it. Simply stated, an extremely low or an extremely high value of a variable will "draw" the mean toward it. Consider the following group of scores: 3, 5, 8, 11, 13. The mean is 40/5 = 8. Now imagine that the score of 13 is changed to 74. The mean has been drawn away from 8 toward the extreme value, 74. In fact, the mean is now 100/5 = 20. It has been increased 2.5-fold by the substitution of one extreme value for one much less extreme!

This is not the case with the median. Rather, it is the middle values of a distribution that contribute to the median. Note that the median remains the same (median = 8) even when a score of 74 is substituted for a score of 13. In this case of extreme positive skew, the mean is not representative of *any* values in the distribution. It is 9 score units away from the nearest data point (20 − 11 = 9). In contrast, the median nicely reflects the four lowest values even when the mean is almost "out of sight."

Just as the sensitivity of the mean to extreme values is one of the weaknesses of that measure, so also is the insensitivity of the median to the extreme value one of its strengths. Since extreme values in one direction or the other affect the mean but not the median, the median should be used as the measure of preference in such situations.

However, the mean is preferred in nearly every other situation. This results from the fact that the mean possesses two unique properties: the sum of the deviations from the mean equals zero and the squared deviations are minimal. Because of these two properties, the mean lends itself ideally for use in more advanced statistical analyses. These properties are illustrated in Table 5.2.

Table 5.2 **Deviations from the Mean, Squared Deviations from the Mean, and Squared Deviations from a Value Other Than the Mean**

X	$(X - \bar{X})$	$(X - \bar{X}^2)$	$(X - 10)$	$(X - 10)^2$
3	$(3 - 8) = -5$	$-5^2 = 25$	$(3 - 10) = -7$	$-7^2 = 49$
5	$(5 - 8) = -3$	$-3^2 = 9$	$(5 - 10) = -5$	$-5^2 = 25$
8	$(8 - 8) = 0$	$0^2 = 0$	$(8 - 10) = -2$	$-2^2 = 4$
11	$(11 - 8) = 3$	$3^2 = 9$	$(11 - 10) = 1$	$1^2 = 1$
13	$(13 - 8) = 5$	$5^2 = 25$	$(13 - 11) = 2$	$2^2 = 4$
Sum	0	68	−11	83

Note that the sum of the second column, where deviations are taken from the mean, equals zero. The sum of the fourth column, where deviations are taken from a value other than the mean, does not equal zero. The sum of the squared deviations (columns 3 and 5) are less when the deviations are taken from the mean (column 3). These properties hold for all distributions, no matter what form they may take.

Finally, the mean is a more stable measure than either the median or the mode. To illustrate, if you were to draw repeated samples from a symmetrical distribution with a single mode (unimodal) and were interested in estimating the population mean, you would find that, over the long run, the sample means would provide greater accuracy (i.e., smaller error) than the median. This is an important advantage for, as we shall see, much work in inferential statistics assumes symmetrical, unimodal population distributions.

Guesstimating the Mean and Median Prior to Performing Any Calculations

When I receive a set of data on which I plan to perform statistical analyses, I always take a few minutes to look over the data and to make guesstimates of statistics such as the mean, the median, and the standard deviation (a measure of dispersion discussed in Chapter 6). These preliminary estimates perform an important service—they alert me to reasonable values of the statistics I am computing, thereby making it possible to detect gross errors in my calculations (such as misplacing a decimal or entering a value of a variable in which I reverse the numbers, such as 82 instead of 28.) These estimates can be extremely accurate, such as when the distribution of values of the variable is relatively symmetrical and unimodal, or inaccurate, such as when the distribution is multimodal and asymmetrical.

The method to use to estimate the mean and median is straightforward and easy to apply. We simply take the arithmetic mean of the lowest and highest values in the distribution. In other words, the estimated mean and median = $(X_{ll} + X_{ul})/2$. Now, if the distribution is symmetrical and unimodal, the mean and median will be identical and at the center of the distribution. The guesstimate will be quite accurate. Even if the symmetry is only approximate, the estimates will be sufficiently close to being accurate so that we will be alerted to gross errors in our calculations if we discover a large disparity between our estimate and our actual calculations. Take the 70 baseline heart measures of Type A subjects (Exercise (d) in Statistics in Action 3.1). The distribution appears to be reasonably symmetrical and only slightly negatively skewed. The mean of the highest and lowest values equals $(42.0 + 101.3)/2 = 71.65$. This is our estimate of both the mean and the median. However, since there appears to be some negative skew, we would expect the computed mean to be a lower value than the computed median. Actual calculations from the grouped frequency distribution yield a mean equal to 72.16 and a median equal to 72.28. These are both reasonably close to our guesstimated mean and median and, as expected, the mean is lower in value than the median. We have no reason to suspect that an error in calculations has been made.

Of course, the guesstimates will not always be this close to actual values. With practice, however, you will be able to revise your guesstimates in the light of the actual distributions and come surprisingly close to the actual values much of the time.

Has the Computer Made the Use of Frequency Distributions and Grouped Frequency Distributions Almost Obsolete?

How often have you heard the computer blamed for some foul-up in the conduct of human affairs? A bank representative tells you that the balance in your savings account was lower than it should have been because of a computer error (Unlikely! It is more likely due to some entry error—the wrong amount was credited or the right amount credited to someone else's account number). Well, statistics is no exception. In the preelectronic calculator and precomputer days, the use of frequency distributions took much of the drudgery out of calculating various statistics. The time and labor required for calculating the mean and median, for example, was reduced substantially. However, with the advent of the computer, enormous quantities of raw data are regularly processed with as much ease as data in the form of frequency distributions. If your sole purpose is to calculate a statistic from data, there is little justification for preparing frequency distributions. However, if you wish to present a mass of data in a visually comprehensible form, there is no substitute for the frequency distribution and the various graphics derived from it. If you don't believe me, try presenting the 140 HR measures in Table 3.6 of the text in the form of an array. Try drawing a frequency curve, a cumulative frequency curve, or a percentage curve without the use of some form of frequency distribution.

Moreover, try reading vital statistics (statistics collected on virtually every aspect of health, disease, life and death, economic well-being, etc.) or census summaries without knowledge of grouped or ungrouped frequency distributions or cumulative frequency and cumulative percentage distributions.

Incidentally, there are numerous statistical packages that are available to owners of the personal computer that will construct frequency distributions from data stored in a data file (cassettes, diskettes, hard disks) or from data entered from the keyboard. Some computers even permit you to define the class boundaries. However, a word of caution: If you ask the computer to calculate statistics from the data you entered, many computers will provide calculations that are based on the *raw data* rather than on the values in the frequency distribution. These calculations are, of course, more accurate but they can cause some dismay to the unwary. Anyone who is subsequently calculating statistics from the frequency distributions will usually obtain somewhat different answers because of grouping errors.

A Worked Example: Calculating the Mean and Median from a Set of Data

In one phase of the study of emotional contrast by Manstead and his associates (1983), the subjects rated their relaxation when humor was presented first versus their relaxation when humor followed horror. The results are shown in Table 5.3.

Table 5.3   Relaxation When Humor Is Presented First Versus Relaxation When Humor Follows Horror*

Humor first		Humor preceded by horror	
Male	Female	Male	Female
21	6	31	10
7	19	9	10
33	16	25	14
11	15	7	7
23	20	6	13
15	37	17	12
20	32	9	6
20	20	28	6
12	12	9	9
15	13	22	15
Sum 177	190	163	102
Mean 17.7	19.0	16.3	10.2
Median 17.5	17.5	13.0	13.25

*Ratings following six scenes of humor were combined to produce a single score for each subject. The lower the score is, the greater is the feeling of relaxation.

Source: Manstead et al., 1983.

1. Construct a bar graph showing the means of the males and females. Place the two categories of humor along the horizontal axis.

2. Combine the male and female scores in each humor category and calculate the mean and median.

3. In your own words, summarize the descriptive statistics that have been calculated.

Answers:

1.

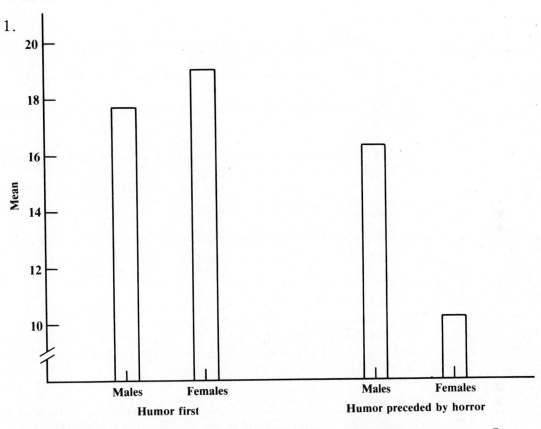

2. Humor first: $\bar{X}$ = 18.35, median = 17.5; humor preceded by horror: $\bar{X}$ = 13.25, median = 10.0.

3. You should have noted that at the descriptive level the results are in the direction that is expected under the emotional contrast hypothesis. The mean relaxation scores are lower (greater relaxation) among both males and females when humor is preceded by horror. This is also true for the combined means: 18.35 versus 13.12. Also, the decrease in the female relaxation means is greater than among the males: 8.8 versus 1.4. Finally, there appears to be very little skew in the humor-first condition, whereas there is more skew (positive) when humor is preceded by horror.

A Worked Example: Calculating the Mean and Median from a Grouped Frequency Distribution

In order to conserve energy and reduce toxic emissions from the exhaust system, federal law has mandated an improvement in the miles-per-gallon ratings. The following table shows miles-per-gallon ratings of 60 automobiles taken at about the same time the laws were enacted.

Class	Midpoint of class	f	fX
28–29	28.5	1	28.5
26–27	26.5	1	26.5
24–25	24.5	2	29.0
22–23	22.5	3	67.5
20–21	20.5	7	143.5
18–19	18.5	6	111.0
16–17	16.5	10	165.0
14–15	14.5	6	87.0
12–13	12.5	10	125.0
10–11	10.5	14	147.0
		$N = 60$	$\Sigma fX = 950$

The following are the steps for calculating the mean of a grouped frequency distribution:

1. Multiply the score at the midpoint of each class by its corresponding frequency. Record in the fX column.

2. Sum the fX column to obtain ΣfX. This sum equals 950.

3. Divide ΣfX by N to obtain the mean:

$$\bar{X} = \Sigma f X/N = 15.83$$

As indicated in the text, the median is merely a special case of a percentile rank. It is the score at the 50th percentile and divides the distribution of score into two halves, with one-half of the cases below and the other half above the median. Using the automobile miles-per-gallon data, the following steps demonstrate the calculation of the median.

Real limits of class	f	Cum f
27.5–29.5	1	60
25.5–27.5	1	59
23.5–25.5	2	58
21.5–23.5	3	56
19.5–21.5	7	53
17.5–19.5	6	46
15.5–17.5	10	40
13.5–15.5	6	30
11.5–13.5	10	24
9.5–11.5	14	14
$\Sigma f = N = 60$		

1. Multiply 50 by N and divide by 100. Thus $(50 \times 60)/100 = 30$.

2. Find the class containing the 30th frequency. This is the class $13.5 - 15.5$.

3. Subtract from the value in step 1 the cumulative frequency at the upper real limit of the *adjacent* lower class. $30 - 124 = 6$

4. Divide the value in step 3 by the frequency within the class and multiply by the i. Thus, $6/6 \times 2$.

5. Add the value in step 4 to the score at the lower real limit of the class containing the frequency, $13.5 + 2.0 = 15.5$

SELECTED EXERCISES

1. The following are number frequency distributions. In the blank spaces, indicate by rank ordering from 1 to 3 which is the most appropriate measure of central tendency, which is second, and which is third. If a measure of central tendency cannot be calculated, mark it with a zero. In some instances, a clear-cut decision is not possible and the issue is debatable. Note that in some distributions the classes are not of equal width.

Distribution A	f	Distribution B	f	Distribution C	f
30–33	2	70 and above	8	0.85–0.89	5
26–29	5	50–69	15	0.80–0.84	8
22–25	8	30–49	17	0.75–0.79	13
18–21	13	20–29	19	0.70–0.74	28
14–17	23	10–19	14	0.65–0.69	10
10–13	12	0–9	8	0.60–0.64	6
6–9	7			0.55–0.59	3
2–5	4			0.50–0.54	0
(−2)–(+1)	1			0.45–0.49	1

Distribution D	f	Distribution E	f	Distribution F	f
above 14,999	20	54–60	2	50,000 & over	12
10,000–14,999	15	47–53	5	35,000–49,999	17
7000–9999	25	40–46	8	25,000–34,999	20
5000–6999	53	33–39	15	20,000–24,999	13
3000–4999	22	26–32	9	15,000–19,999	12
2000–2999	8	19–25	9	12,500–14,999	6
1000–1999	4			7500–12,499	11
0–999	3			2500–7499	7
				0–2499	2

Distribution	Mean	Median	Mode	Reason
A				
B				
C				
D				
E				
F				

2. For each of the distributions in Selected Exercise 1, find the three measures of central tendency, except in instances in which one or more measures may not be calculated.

3. Find the mean and median for the four distributions constructed in Selected Exercise 2, Chapter 3 of the Study Guide (p. 38).

4. The number of students graduating from five high schools within a school district and their mean verbal SAT scores were as follows:

School	1	2	3	4	5
Mean	509.7	489.33	516.89	503.44	519.76
N	225	180	211	162	195

Find the weighted mean SAT score for the entire school district.

Answers:

1.

Distribution	Mean	Median	Mode	Reason
A	1	2	3	Fairly symmetrical
B	0	1	2	Indeterminate value
C	1	2	3	Symmetrical
D	0	1	2	Highly skewed
E	1	2	3	Fairly symmetrical
F	0	1	2	Indeterminate value

2.

Distribution	Sum	Mean	N	Median	Mode
A	1202.5	16.03	75	18.85	15.5
B	—	—	81	29.24	24.5
C	53.68	0.7254	74	0.7254	0.72
D	—	—	150	6,433.46	5,999.5
E	1707	35.56	48	33.20	36.0
F	—	—	100	24,614.9	22,549.5

3.

Distribution	Mean	Median
$i = 1$	20.81	19.12
$i = 3$	20.78	18.94
$i = 5$	21.00	19.33
$i = 8$	20.90	19.83

4. $\sum X = 494736$, $N = 973$, $\bar{X} = 508.465$

SELF-QUIZ: TRUE-FALSE

Circle T or F.

T F   1. The term average is clear-cut and unambiguous.

T F   2. A frequency distribution permits direct quantitative statements describing various statistics of that distribution.

T F   3. Data frequently cluster around a central value that is between the two extreme values of the variable under study.

T F   4. Measures of central tendency commonly permit us to reduce a mass of data to a single quantitative value.

T F   5. The median of the scores 6, 5, 8, 4, 7 is 5.

T F   6. The mean of the scores 6, 5, 8, 4, 7 is 5.

T F   7. The median is like the fulcrum of a teeter board, balancing the values on either side of it.

T F   8. The sum of deviations of scores from the mean of a distribution equals 0.

T F   9. Suppose there are two otherwise identical data sets with one extremely high score in one of the sets. The median of the data set with the high score will be larger than the mean.

T   F   10. Suppose there are two otherwise identical data sets with one extremely high score in one of the sets. The median of the data set with the high score will be larger than the median of the data set without the high score.

T   F   11. The sum of the squared deviations about the mean is minimal.

T   F   12. For symmetrical distributions, the mean may be regarded as a value above and below where one-half of the frequencies lie.

T   F   13. The median is the score at the 50th percentile.

T   F   14. The median is generally the preferred measure of central tendency.

T   F   15. If we were to take repeated samples from a given population, the median would usually fluctuate less than either the mean or the mode.

T   F   16. The mean underestimates the income obtained by most families.

T   F   17. The mode is preferred whenever we desire a precise and useful estimate of central tendency.

T   F   18. In a positively skewed distribution, the mean will be to the right of the median.

T   F   19. In purchasing meat for a family of five, you paid the following prices per pound of chuck steak: $1.05, $1.55, $0.83, and $1.33. The mean cost per pound for these purchases was $1.19.

T   F   20. If the Ns are equal when determining the weighted mean of a set of values, we may use the same formula as for the arithmetic mean.

T   F   21. A brokerage house sells varying numbers of shares of common stock at $2, $3, and $4. The mean cost per share is $3.

T   F   22. If two extreme scores are added to a distribution—one exactly 20 units above and one 20 units below the mean—the mean will be drawn toward the higher score.

T   F   23. The following collection of values constitutes an array: 25, 11, 13, 42, 12, 9, 16.

T   F   24. If Andy were to examine the values in the preceding example by finding the sum of the squared value divided by N, he would be using the least squares method of locating the mean.

T   F   25. The mean and mode are the same in all symmetrical distributions.

T   F   26. The formula for the median is a modification of the formula for finding the percentile rank of a score of 50.

T   F   27. When a distribution contains classes with indeterminate boundary values, the mean is preferable if the distribution is symmetrical.

Answers: (1) F; (2) F; (3) T; (4) T; (5) F; (6) F; (7) F; (8) T; (9) F; (10) F; (11) T; (12) T; (13) T; (14) F; (15) F; (16) F; (17) F; (18) T; (19) T; (20) T; (21) F; (22) F; (23) F; (24) F; (25) F; (26) F; (27) F.

SELF-TEST: MULTIPLE-CHOICE

1. Given that a distribution of scores yields a mean of 40, a median of 38, and a mode of 36, if we added 12 points to each score, what would the new median be?

   a) 40   b) 42   c) 50   d) 52   e) insufficient information to answer

2. What is the new mean?

   a) 40   b) 42   c) 50   d) 52   e) insufficient information to answer

3. If 12 points are subtracted from each score, the mean is:

   a) unchanged   b) increased by 12 points   c) decreased by 12 points
   d) equal to $(\bar{X}/12) \cdot N$ e) to answer, value of mean must be known

4. For a given distribution, the mode is 68, the median is 62, and the mean is 56. This distribution is:

   a) normal   b) symmetrical   c) positively skewed   d) negatively skewed
   e) leptokurtic

5. $N\bar{X}$ is equal to:

   a) $(X_1 + \bar{X}) + (X_2 + \bar{X}) + \cdots (X_N + \bar{X})$
   b) $\sum X$
   c) $\sum X/N$
   d) $X_1 + X_2 + \cdots X_N$
   e) none of the above

6. $\sum(X - \bar{X})$ equals:

   a) $\sum X - \sum X$ b) 0   c) $\sum X - N\bar{X}$
   d) all of the above   e) none of the above

7. $\sum X$ equals:

   a) $\bar{X}/N$ b) $\bar{X} \cdot N$ c) $N/\bar{X}$ d) all of the above   e) none of the above

8. In the formula median $= X_{ll} + i(N/2 - \text{cum } f_{ll})/N$, which of the following does cum f_{ll} stand for?

   a) the frequency in the class containing the median
   b) the total frequency up to the lower real limit of the class containing the median
   c) the total frequency up to the upper real limit of the class containing the median
   d) the class containing the mode
   e) none of the above

Multiple-Choice Problems 9 through 13 refer to the following frequency distributions:

Class	Group A, f	Group B, f	Group C, f
95–99	6	12	7
90–94	13	14	9
85–89	11	6	3
80–84	7	2	7
75–79	2	1	6

9. Each group has the same:

   a) N b) mean   c) median   d) mode   e) skew

10. The median of grouped frequency distribution B is in the class:

   a) 95–99   b) 90–94   c) 85–89   d) 80–84   e) 75–79

11. The mean of grouped frequency distribution A is in the class:

   a) 95–99   b) 90–94   c) 85–89   d) 80–84   e) 75–79

12. The median of grouped frequency distribution C is:

    a) 84.5   b) 85   c) 87   d) 89   e) 89.5

13. The distribution of scores for Group B is:

    a) symmetrical   b) normal   c) positively skewed   d) negatively skewed
    e) bimodal

14. When an *odd* number of scores is arranged in an array, the median is:

    a) the score with the greatest frequency
    b) the middle score
    c) the mean of the two middle scores
    d) the mean of the highest and lowest scores
    e) cannot be determined without additional information

15. When an *even* number of scores is arranged in an array, the median is:

    a) the score with the greatest frequency   b) the middle score   c) the mean
    of the two middle scores   d) the mean of the highest and lowest scores
    e) always equal to the mean

16. A group of 20 students obtained a mean score of 70 on a quiz. A second
    group of 30 students obtained a mean of 80 on the same quiz. The overall
    mean for the 50 students was:

    a) 70   b) 74   c) 75   d) 76   e) 80

17. Which measure of central tendency yields the most prosperous picture of
    income in the United States?

    a) mean   b) mode   c) median   d) weighted mean   e) all the same

18. The most frequently occurring score is:

    a) any measure of central tendency   b) mean   c) median   d) mode
    e) none of the above

19. If the median and the mean are equal, you know that:

    a) the distribution is symmetrical
    b) the distribution is skewed
    c) the distribution is normal
    d) the mode is at the center of the distribution
    e) the distribution is positively skewed

20. If the mean and median are unequal, you know that:

    a) the distribution is symmetrical
    b) the distribution is skewed
    c) the distribution is normal
    d) the mode is at the center of the distribution
    e) the distribution is positively skewed

21. The mean, median, and mode are all measures of:

    a) the midpoint of the distribution   b) the most frequent score
    c) percentile ranks   d) variability   e) none of the above

22. It is usually possible to see at a glance when looking at a frequency distribution:

    a) the mean   b) the median   c) the mode   d) the midpoints of the classes
    e) the score at the 50th percentile

23. In what type of distribution might the mean be at the 60th percentile?

a) positively skewed   b) negatively skewed   c) normal   d) leptokurtic
e) U distribution

24. In what type of distribution might the mean be at the 40th percentile?

a) positively skewed   b) negatively skewed   c) normal   d) leptokurtic
e) U distribution

25. Which of the following is most likely to be the median of a positively skewed distribution with a mean of 65 and a mode of 57?

a) 40   b) 47   c) 60   d) 65   e) 73

26. Which of the following is most likely to be the median of a negatively skewed distribution with a mean of 57 and a mode of 65?

a) 40   b) 47   c) 60   d) 76   e) 79

27. Given that the mean for 60 students is 75 and the mean for the 40 women in the class is 79, what is the mean for the 20 men?

a) 67   b) 71   c) 75   d) 77   e) 79

28. In a bell-shaped distribution of scores:

a) the mean, median, and mode are the same
b) the mean is usually higher than the median
c) the median is usually higher than the mean
d) the mean and median are the same but the mode is different
e) none of the above

29. Which of the following constitutes the definition of the mean?

a) the sum of all scores divided by the number of scores
b) the point in a distribution about which the summed deviations equal 0
c) the point in a distribution about which the sum of squared deviations is minimal
d) all of the above   e) (a) and (b) but not (c)

30. If 200 shares of common stock are purchased as follows—50 at $2.00 per share, 75 at $2.50, 25 at $3.00, and 50 at $1.50—the break-even price is approximately:

a) 2.25   b) 2.19   c) 4.45   d) 4.38   e) 4.45 dollars per share

Answers: (1) c; (2) d; (3) c; (4) d; (5) b; (6) d; (7) b; (8) b; (9) d; (10) b;
(11) c; (12) e; (13) d; (14) b; (15) c; (16) d; (17) a; (18) d; (19) a; (20) b;
(21) e; (22) c; (23) a; (24) b; (25) c; (26) c; (27) a; (28) a; (29) d; (30) b.

Formulas for Use in Chapter 5 of the Text

$\bar{X} = \sum X/N$ the mean of a set of scores

$\bar{X} = \sum fX/N$ the mean of an ungrouped or grouped frequency distribution

$\sum X = \bar{X}/N$ the sum of scores when mean and N are known

$N = \sum X/\bar{X}$ the sample size (N) when mean and $\sum X$ are known

$\bar{X}_w = \sum f\bar{X})/N$ the weighted mean

Median $= X_{ll} + i(N/2 - \text{cum } f_{ll})/N$ the median of a grouped frequency distribution

Median $= X_{ll} + (N/2 - \text{cum } f_{ll})/N$ the median of an ungrouped frequency distribution

Measures of Dispersion

BEHAVIORAL OBJECTIVES

Conceptual Objectives

1. Describe the purpose of measuring the dispersion or variability of scores about the measure of central tendency. State the relationship between the shape of the distribution and the measurement of dispersion.

2. Define the crude range of a scale of scores. Explain the usefulness of the crude range.

3. Define and calculate the semi-interquartile range. Identify its shortcomings and its advantages.

4. Specify, both in words and in algebraic notation, the mean deviation and the sum of all the deviations from the mean without regard to sign. Explain the advantages and disadvantages of using the mean deviation.

5. Identify the purposes served by the standard deviation. In words and algebraic form, define the standard deviation and the variance. Explain the exact relationship between the variance and the standard deviation.

6. Use the crude range as the basis for guesstimating the mean and the standard deviation of a data set.

CHAPTER REVIEW

As we saw in Chapter 5, there are two aspects of data sets on which statisticians concentrate when quantitatively describing these sets. We have already discussed the first of these—the central value around which the scores in the distribution tend to cluster. In Chapter 6, we examine a second set of measures that are used to describe the variability of scores, or the extent to which the scores are dispersed, or scattered, about some central value. The first measure is known as central tendency, the second is known as dispersion or variability.

A researcher can communicate a wealth of information about a data set by citing the values of only two quantities—central tendency and dispersion. However, knowing one without its companion measure does not provide a satisfactory basis for describing the salient features of a distribution of scores.

Just as there are several different measures for describing central tendency, so there are also several measures for quantifying variability; each is a measure of dispersion. Which measure is used for any particular application depends on a variety of factors, including the symmetry of the distribution and the assumptions underlying the use and interpretation of each measure.

The simplest and most straightforward measure of variability is the crude range, defined as the scale distance between the largest and smallest score. Thus, if the maximum score in a distribution is 163 and the minimum value is 24, the range is $163 - 24 = 139$. As the adjective "crude" implies, the range is the least refined of all the measures of dispersion. Since it is based on only the two most extreme scores, one at each end of the distribution, it can be "the victim" of one or two extremely deviant values. Its marked instability makes the range one of the least useful measures of dispersion. However, as we will see later in the chapter, it can serve as a starting point for guesstimating other measures of variability as well as measures of central tendency.

A step above the range in sophistication is the semi-interquartile range. Although not so unstable as the crude range, it is not usually the measure of preference for quantifying variability. To calculate the semi-interquartile range, you subtract the score as the 25th percentile from the score at the 75th percentile and divide by 2. Symbolically, the score at the 75th percentile is known as Q_3 and the score at the 25th percentile is Q_1 (the median is Q_2). If $Q_3 = 112$ and $Q_1 = 36$, the semi-interquartile range is 38. Note that 50% of all cases in a distribution lie between Q_3 and Q_1 and is known as the interquartile range. In other words,when you divide the interquartile range by 2, you obtain the semi-interquartile range.

In Chapter 5, we discussed the sum of the deviations from the mean, $\Sigma(X - \bar{X})$. You should also recall a defining property of the mean is that the sum of these deviations must equal zero. In order to understand the *mean deviation* measure of dispersion, you should keep these facts in mind. Just remember that the deviations from the mean $(X - \bar{X})$ are the basis for the calculation of the mean deviation.

The mean deviation makes use of the absolute values of the deviations. As you may recall from algebra, the absolute value of a number is its positive value. Thus, the absolute value of a positive number (or zero) is the number itself. The absolute value of a negative number is its positive value. The notation $|8|$ instructs you to take the absolute value of 8, which is, naturally, 8. The absolute values of $|-10|$ and $|0|$ are 10 and 0, respectively.

The formula for the mean deviation (MD) combines the concepts of absolute values and deviations from the mean:

$$MD = \frac{\Sigma(|X - \bar{X}|)}{N}$$

In other words, the mean deviation is the sum of the absolute values of the deviations from the mean divided by N.

The mean deviation is useful for comparing the dispersion or variability of two different distributions. The greater the mean deviation is, the greater is the dispersion about the mean. The mean of the values 4, 5, and 6 equals 5 with a mean deviation of 2/3 or 0.67. In contrast, the values 3, 5, and 7 have the same mean (5) but the mean deviation is 4/3 = 1.33. The second set of values contains scores with greater deviations from the mean. Thus, its variability is greater.

In the mean deviation method of quantifying dispersion, we used absolute values to eliminate negative values, assuring that the sum of these absolute deviations from the mean will not equal zero. In fact, the only time that the sum of the absolute values of the deviations will equal zero is when all the values in a data set are identical. In that event, the mean deviation will also equal zero.

In calculating the variance and standard deviations, we again use the deviations from the mean, but this time a different technique is used to avoid a zero value. Rather than using the absolute value of the deviations, we square the quantity $(X - \bar{X})$ for each score in the distribution. Then these squared deviations are added to obtain the quantity $\Sigma(X - \bar{X})^2$. When you divide this quantity by the sample size (N), you obtain the variance of the distribution:

$$s^2 = \frac{\Sigma(X - \bar{X})^2}{N}$$

Stated verbally, the variance is the sum of the squared deviations from the mean divided by N. The standard deviation is closely related to the variance. In fact, once you have calculated the variance of a data set, you need do no more than take the square root of the variance to obtain the standard deviation. Thus

$$s = \frac{\Sigma(X - \bar{X})^2}{N}$$

or

$$s = \sqrt{s^2}$$

The quantity $\Sigma(X - \bar{X})^2$ is known as the sum of squares and, for convenience, may be replaced by SS in the formula for both the variance and the standard deviation. Thus

$$s^2 = SS/N$$

and

$$s = \sqrt{SS/N}$$

To provide practice calculating the variance and standard deviation of a data set, we provide the following table. Fill in the blanks as indicated. Find the values to the nearest hundredth. *Note:* The mean must be calculated first. (Answers are shown in the two tables that follow.)

Score (X)	Deviation $(X - \bar{X})$	Deviation squared $(X - \bar{X})^2$
39	10	100
31	___	___
30	1	___
26	___	9
25	−4	___
23	___	___
$\Sigma X = 174$		$\Sigma X^2 = $ ___
$\bar{X} = $ ___ SS = ___	$s^2 = $ ___ and $s = $ ___	

From the textbook, you know that of the five measures of dispersion we have discussed thus far, the standard deviation (and the variance) will be the most widely used and most valuable. The many reasons for this will become apparent to you in Chapter 7. There are many different formulas that can be used to calculate the variance and standard deviation. The following is a raw score formula for finding the sum of squares (SS). The use of the raw score formula

Completed Table of Calculations of s and s^2
Mean Deviation Method

Score (X)	Deviation $(X - \bar{X})$	Deviation squared $(X - \bar{X})^2$
39	10	100
31	2	4
30	1	1
26	−3	9
25	−4	16
23	−6	36
$\Sigma X = 174$		$\Sigma(X - \bar{X})^2 = 166$
$\bar{X} = 174/6 = 29$		SS = 166
$s^2 = 166/6 = 27.67$ and $s = 5.26$		

eliminates the necessity of finding the deviation of each score from the mean, squaring, and then summing these squared deviations. Thus, this formula is easier to use with most of the data sets you will encounter:

$$SS = \sum X^2 - (\sum X)^2/N$$

The following demonstrates the calculations that are necessary to find s^2 and s, using the raw score method.

Score (X)	Score squared X^2
39	1521
31	961
30	900
26	676
25	625
23	529
$\sum X = 174$	$\sum X^2 = 5221$

Step 1:   Sum the values of the variable to obtain $\sum X = 174$ and count $N = 6$.

Step 2:   Square each value of the variable and then sum these squared values. Thus, $\sum X^2 = 5212$.

Step 3:   Calculate the sum of squares: $SS = \sum X^2 - (\sum X)^2/N$
$$= 5212 - (174)^2/6$$
$$= 5212 - 5046$$
$$= 166$$

Step 4:   Find the sample variance by dividing SS by N: $166/6 = 27.67$.

Step 5:   Find the sample standard deviation by extracting the square root of the variance. Thus, $s = \sqrt{27.67} = 5.26$.

Note that both of these statistics are the same as when obtained via the mean deviation method. However, the raw score method is more readily obtained.

Using Observational Techniques to Guesstimate the Mean and Standard Deviation

As noted earlier, one way to avoid unknowingly accepting gross errors when calculating a statistic is to guess the approximate value prior to conducting the formal statistical analysis. There are two methods that provide reasonably accurate approximations on most occasions. One uses the crude range and the other employs the score at the 16th percentile. Let's look first at the guesstimate that uses the crude range.

To begin with, simply add together the score values at each extreme of the distribution and divide by 2 if you want a rough-and-ready approximation to the mean. If the distribution of scores is relatively symmetrical, the approximation will often be remarkably close. For example, the scores made by Type B subjects on a standard challenging task (Table 6.2 of the text) appear to be reasonably symmetrical. The high score was 10 and the low score was 5. The guesstimated mean, then, is $(5 + 10)/2 = 7.5$. The actual mean was 7.75. In contrast, in the study of emotional contrast (Statistics in Action 5.1 in the text) when horror was preceded by humor, the two extreme values for males were 20 and 36. The guesstimated mean is $56/2 = 28$. The actual mean was 32.9. By inspection, it is obvious that there was a pile-up of scores at the high end of the distribution. Since the mean is drawn in the direction of this cluster of extreme values, we would expect the mean to be higher than 28. If we obtain a mean lower than 28, we can be quite sure an error has been made.

It is well known that the ratio of the crude range to the standard deviation is rarely smaller than 2 or larger than 6. In my experience, a ratio of 3 provides a fairly reasonable guesstimate of the standard deviation for most distributions. In other words, if you divide the crude range by 3, you'll usually come fairly close to the actual standard deviation. This is particularly the case when the distribution is symmetrical and bell-shaped. If there is a marked lack of symmetry, your guesstimate will usually be too small. Referring back to Table 6.2 of the text, the crude range was 5. Dividing by 3, we obtain a guesstimated s equal to 1.67. The actual standard deviation was 1.55. In the study of seasonal affective disorder (Case Example 6.1), the range of scores on the Hamilton Rating Scale $(32.5 - 15)$ was 17.5. Dividing by 3, we obtain a guesstimate of 5.83. The actual value was 5.74.

The second guesstimating technique makes use of the knowledge that for bell-shaped distributions the distance between the mean and the score at the 16th or 84th percentile equals the standard deviation. The reasons for this will be discussed in Chapter 7 of the text. Suffice it to note for the moment that if you have a guesstimated mean and a guesstimated score at the 16th percentile and a bell-shaped distribution, you merely subtract the latter from the mean to obtain a reasonably good approximation to the standard deviation. If you have a score at the 84th percentile, subtract the mean from that score in order to approximate the standard deviation. To illustrate, in Table 6.5 of the text, the score at the 16th percentile is approximately 6 (i.e., multiplying $N = 20$ by .16, we know that the 3.2nd score is at the 16th percentile), since the 3.2nd score is approximately halfway between 5.5 and 6.5. The mean is known to be 7.75. Thus, the guesstimated standard deviation is $7.75 - 6 = 1.75$. The actual standard deviation is 1.55.

Converting from Variance Calculation Using $N - 1$ in the Denominator to Variance Calculation Using N in the Denominator

As noted in the text, occasionally a calculator or computer software programmed to calculate various statistics (usually designated as *statistical* calculator or software) may use $N - 1$ in the denominator when calculating s^2 and s. If you have one of these calculators, simply multiply the variance given by your calculator by $(N - 1)/N$. To illustrate, in the example appearing in the preceding section, the variance, using $N - 1$ in the denominator, equals 33.2. When multiplying by $(N - 1)/N$, we obtain $(33.2 \times 5)/6 = 27.67$, which is what we obtained when using N in the denominator. The standard deviation is the square root of this quantity, or 5.26.

SELECTED EXERCISES

1. It was pointed out in the text that the ratio of the standard deviation to the crude range is rarely smaller than 2 or greater than 6. Calculate the standard deviation of the following distributions, all of which have a crude range of 4 and an N of 9. Determine the ratio of range to the standard deviation in each case. Generalize: In which type of distribution is the ratio likely to approach 6 and in what form of distribution does it approach 2?

X	f_a	f_b	f_c	f_d	f_e
6	1	1	2	3	4
5	1	2	2	1	0
4	5	3	1	1	1
3	1	2	2	1	0
2	1	1	2	3	4

2. Given the following observations, calculate the variance and the standard deviation.

22	16	13	11	9	5
22	15	13	11	9	4
20	15	13	11	9	4
20	15	13	11	9	3
19	14	12	10	8	1
18	14	12	10	8	1
18	14	12	10	7	
17	14	12	10	7	
17	13	11	9	7	
17	13	11	9	6	
16	13	11	9	6	

3. Determine the ratio between the standard deviation and range for the data in Selected Exercise 2.

4. One way of approximating the standard deviation for data that are approximately normally distributed is to find the mean and the score at the 16th percentile. By subtracting the later from the former, you can obtain a reasonable approximation to s. The reason for this is made clear in Chapter 7. Employing the data in Selected Exercise 2, determine if the mean minus the score at the 16th percentile approximates the standard deviation.

5. Another way of approximating the standard deviation is to find the score at the 84th percentile and then subtract the mean from it. Employing the data in Selected Exercise 2, determine if the score at the 84th percentile minus the mean approximates the standard deviation.

6. Given the following 50 observations, calculate the variance and the standard deviation.

18	37	15	31	19	31
33	32	30	34	26	17
28	15	39	20	10	25
29	39	5	24	36	33
22	30	13	23	26	35
23	18	20	26	25	
14	5	41	12	27	
19	25	29	15	51	
11	38	21	27	20	

7. An astronomer studying a group of stars very close to the earth, relatively speaking, of course, approximated their distances from our planet. Find the mean distance and the standard deviation of the distribution using the mean deviation formula.

Star	Distance (in light years)
1	8.6
2	7.2
3	10.1
4	4.3
5	5.0
6	4.8
7	9.0
8	7.6
9	7.0
10	8.6

Answers:

1.

Distribution	Standard deviation	Range/standard deviation
a	1.05	3.81
b	1.15	3.48
c	1.49	2.68
d	1.70	2.35
e	1.89	2.12

2. $\sum X = 709$; $\sum X^2 = 7603$; $N = 61$; $s^2 = 22.33$; $s = 4.73$.

3. Ratio of range to standard deviation = $21/4.73 = 4.44$.

4. In Selected Exercise 2, $\bar{X} = 11.62$ and the score at the 16th percentile is 7.09. The difference is $11.62 - 7.09 = 4.53$. This is quite a good approximation to the obtained $s = 4.73$. The approximation will not be as good for skewed and/or nonnormal distributions.

5. In Selected Exercise 2, $\bar{X} = 11.62$ and the score at the 84th percentile is 16.58. The absolute difference (i.e., without regard to sign) is: $16.58 - 11.62 = 4.96$. Again, this is quite a good approximation to obtained $s = 4.73$.

6. $\sum X = 1242$; $(\sum X)^2 = 1,542,564$; $N = 50$; $\sum X^2 = 35,438$; SS = 4586.74; $s^2 = 91.73$; $s = 9.58$.

7. $\sum X = 72.2$; $\sum X^2 = 556.06$; SS = 34.7761; $N = 10$; $s^2 = 3.478$; $s = 1.86$.

SELF-QUIZ: TRUE-FALSE

Circle T or F.

T F   1. A given score above the mean is likely to be "higher" when the scores are compactly distributed about the mean than when widely dispersed about the mean.

T F   2. A companion to a measure of central tendency is a measure that expresses the degree of dispersion of scores about the central tendency.

T F   3. Using the median as a norm, fully half the population is overweight.

T F   4. The most useful measure of dispersion is the semi-interquartile range.

T F   5. The absolute value of –9 is 9.

T F   6. If a normal distribution is assumed, the standard deviation permits a precise interpretation of scores within a distribution.

T F   7. The variance is always larger than the standard deviation.

T F   8. The standard deviation and the mean are both members of a mathematical system that permits their use in more advanced statistical considerations.

T F   9. The standard deviation of 88, 89, 90 is greater than the standard deviation of 0, 1, 2.

T F   10. If the variance of a distribution is 4, the standard deviation is 16.

T F   11. The expression $\sqrt{\dfrac{SS}{N}}$ is equivalent to $\sqrt{\dfrac{\sum(X - \bar{X})^2}{N}}$.

T F   12. The sum of squares of deviations from the mean may be represented symbolically as $\sum X^2$.

T F   13. The sum of squares may be defined as $\sum X^2 - \dfrac{(\sum X)^2}{N}$.

T F   14. If $\sum X^2 = 20$ and $N = 5$, we know the variance is equal to 4.

T F   15. The scores, 0, 2, 4 have a greater standard deviation than the scores 88, 89, 90.

T F   16. The range decreases as we increase N.

T F   17. When studying dispersion, we are interested in an index of variability that indicates the *distance* along a scale of scores.

T F   18. Of all the measures of dispersion, the semi-interquartile range is the least stable.

T F   19. The main similarity between the mean deviation measure and the variance is that neither takes into account the zero value of the deviations from the mean.

T F   20. As a measure of variability, the worth of the mean deviation is as a basis for comparing distributions.

T F   21. The symbol for the standard deviation is s, and for the variance, s^2.

T F   22. Unlike most other measures of dispersion, the standard deviation allows for the precise interpretation of scores when the underlying distribution is normal.

T F   23. The variance is the square root value of the standard deviation.

T F   24. The concept of the sum of squares can be represented algebraically by the term $\sum(X - \bar{X})^2$.

T F   25. When the mean is a fractional value, the mean deviation formula for calculating the standard deviation can prove to be unwieldly.

Answers: (1) T; (2) T; (3) T; (4) F; (5) T; (6) T; (7) F; (8) T; (9) F; (10) F; (11) T; (12) F; (13) T; (14) F; (15) T; (16) F; (17) T; (18) F; (19) F; (20) T; (21) T; (22) T; (23) F; (24) T; (25) T.

SELF-TEST: MULTIPLE-CHOICE

1. Which statistic does not belong with the group?

   a) range   b) mean   c) semi-interquartile range   d) standard deviation
   e) mean deviation

2. $(\Sigma X)^2$ is equal to:

a) ΣX^2 b) $X_1^2 + X_2^2 + \cdots + X_N^2$ c) NX^2
d) all of the above   e) none of the above

3. Which of the following groups of scores exhibits the *least* variability?

a) 2, 4, 6, 8, 10, 12   b) 2, 3, 4, 10, 11, 12   c) 2, 6, 7, 7, 8, 12
d) 2, 2, 3, 11, 12, 12   e) all the same

4. Which group of scores exhibits the *most* variability?

a) 2, 4, 6, 8, 10, 12   b) 2, 3, 4, 10, 11, 12   c) 2, 6, 7, 7, 8, 12
d) 2, 2, 3, 11, 12, 12   e) all the same

5. Distributions A and B have the same mean and range. The standard deviation of Distribution A is 15 and of Distribution B is 5. We may conclude that:

a) the scores in Distribution A are grouped closer to the mean than are the scores in Distribution B
b) the scores in Distribution B are grouped closer to the mean than are the scores in Distribution A
c) there are three times as many scores from −1 standard deviation to +1 standard deviation in Distribution A
d) there are one-third as many scores from −1 standard deviation to +1 standard deviation in Distribution A
e) cannot say unless we know the value of the mean

6. Reducing the frequency from the middle of a distribution will:

a) increase the standard deviation
b) reduce the standard deviation
c) not affect the standard deviation
d) depend on the N
e) depend on the range

7. Reducing the frequency from the tails of a distribution will:

a) increase the standard deviation
b) reduce the standard deviation
c) not affect the standard deviation
d) depend on the N
e) depend on the range

8. Which of the following measures of variability is *not* dependent on the exact value of each score?

a) range   b) mean deviation   c) standard deviation   d) variance
e) none of the above

9. In a certain frequency distribution with a mean of 50 and a standard deviation of 10, six scores of 10 are added to the distribution. The recomputed value of s will be:

a) greater than the original s
b) less than the original s
c) unchanged
d) dependent on the N in the original group
e) 60% of the original s

Questions 10 through 13 refer to the following information: The mean score of 500 students on a statistics exam is 45 and the s is 5.

10. If 2 points were added to each of the 500 scores, the new s would be:

a) 2.5   b) 3   c) 5   d) 5 + (2/500)   e) 7

11. If 2 points were subtracted from each of the 500 scores, the new *s* would be:

    a) 2.5   b) 3   c) 5   d) 5 – (2/500)   e) 7

12. If each score were doubled, the new *s* would be:

    a) 2.5   b) 5   c) 7   d) 10   e) 5 + (2/500)

13. If each score were divided in half, the new *s* would be:

    a) 2.5   b) 5   c) 7   d) 10   e) 5 + (2/500)

14. A test given to three groups yielded the same mean. There were 10 subjects in Group A, 20 in Group B, and 30 in Group C. Which of the following statements is true concerning the standard deviations?

    a) *s* of Group C is the largest
    b) *s* of Group A is the largest
    c) *s* of Group C is 3 times as large as *s* of Group A
    d) all the *s*'s are equal
    e) none of the above

15. A child is born in the Ztat family every year for seven consecutive years. The *s* of the ages of the Ztat children:

    a) equals 1   b) equals 2   c) equals 4   d) equals 7
    e) impossible to calculate without knowing the actual ages

16. The mean score of 30 students on an exam is 47 and the *s* is 0. The distribution of scores:

    a) is mesokurtic   b) is normal   c) contains many negative scores
    d) all students scored 47   e) not enough information

17. Each of the following frequency distributions has the same top and bottom scores. Which one has the largest standard deviation?

    a) the *U* distribution
    b) the positively skewed distribution
    c) the negatively skewed distribution
    d) the leptokurtic distribution
    e) the normal distribution

18. On a statistics examination, an instructor finds that *s* = 14.0 in an undergraduate class and 7.0 in a graduate seminar. Both groups have the same number of students. Which of the following statements is warranted?

    a) the distribution is more likely to be normal among undergraduate than graduate students
    b) undergraduates are more heterogeneous in performance than graduate students
    c) the average performance for graduates is higher than for undergraduates
    d) undergraduates performed better than graduate students
    e) cannot say unless we know the means

19. $\Sigma(X - \bar{X})^2$ equals:

    a) $\Sigma X^2 - \dfrac{(\Sigma X)^2}{N}$ b) SS   c) $\Sigma X^2 - N\left(\dfrac{\Sigma X}{N}\right)^2$

    d) all of the above   e) none of the above

20. s^2 equals:

    a) $\dfrac{SS}{N}$ b) $\dfrac{\Sigma X^2}{N}$ c) $\sqrt{\dfrac{\Sigma X^2}{N}}$ d) $\sqrt{\dfrac{SS}{N}}$ e) $\sqrt{\dfrac{\Sigma X^2 - (\Sigma X)^2}{N}}$

21. Which of the following measures is generally our most useful measure of dispersion?

    a) the range   b) the semi-interquartile range   c) the standard deviation   d) the mean deviation   e) all are about equally useful

22. Which measure of dispersion reflects only the two most extreme scores in a distribution?

    a) the mean deviation   b) the standard deviation   c) the semi-interquartile range   d) the range   e) the variance

23. Which measure of dispersion is defined as the sum of the deviations from the mean divided by N?

    a) the range   b) the mean deviation   c) the standard deviation   d) the semi-interquartile range   e) no measure of dispersion is so defined

24. Distribution W has a standard deviation of 4. If a score of 20 has a corresponding z-score of -2, what is the mean of the distribution?

    a) 12   b) -28 c) -12 d) 28
    e) impossible to calculate from the preceding information

25. A given distribution has a mean of 60. A score of 75 has a corresponding z of $+2.00$. The standard deviation is:

    a) 7.5   b) -2.00 c) 1.25   d) 2.00   e) none of the above

26. From a sample of ten baseball players the mean batting average is .287 and the median score is .278 with a standard deviation of .1. What is the index of skew?

    a) $-.27$ b) 2.7   c) -2.7 d) .27

27. What is the range of the following scores: 8, 26, 10, 36, 4, 15?

    a) 7   b) 11   c) 32   d) 28

28. The scores at various percentile ranks were 10th, $X = 20$; 15th, $X = 25$; 25th, $X = 30$; 35th, $X = 35$; 50th, $X = 40$; 65th, $X = 45$; 75th, $X = 50$; 85th, $X = 55$; 90th, $X = 60$. The semi-interquartile range is:

    a) 10   b) 20   c) 25   d) 12.5

29. The mean deviation of the following scores—2, 6, 10—is:

    a) 1.33   b) 4.00   c) 6.00   d) 2.67

30. The standard deviation of the following scores—2, 6, 10—is:

    a) 4.00   b) 1.63   c) 16   d) 3.27

31. The variance of the following scores—2, 5, 8, 11—is:

    a) 11.25   b) 15.00   c) 214   d) 26.00

32. The sum of squares of the following scores—5, 7, 9, 11—is:

    a) 276   b) 1024   c) 8   d) 20

33. The sum of squares of the following scores—1, 3, 3, 3, 5—is:

    a) 4   b) 8   c) 53   d) 15

Answers: (1) b; (2) e; (3) c; (4) d; (5) b; (6) a; (7) b; (8) a; (9) a; (10) c; (11) c; (12) d; (13) a; (14) e; (15) b; (16) d; (17) a; (18) b; (19) d; (20) a; (21) c; (22) d; (23) e; (24) d; (25) a; (26) d; (27) c; (28) a; (29) d; (30) d; (31) a; (32) d; (33) b.

The Standard Deviation and the Standard Normal Distribution

BEHAVIORAL OBJECTIVES

Conceptual Objectives

1. State the function of standard scores and define them in words and algebraic notation.

2. Use standard scores to compare the values of two different variables.

3. State the four functions of z-scores in words and in symbolic notation.

4. Describe the properties of the standard normal distribution and specify the relationship between the standard deviation and the area of the standard normal distribution.

5. Explain the function of Table A, located in the appendix to the text, in determining the area between two scores of the standard normal distribution. Specify the assumptions underlying the use of this table.

6. When the mean is used as the basis for predicting scores, describe how the precision of the estimate is influenced by the magnitude of the standard deviation.

Procedural Objectives

1. Given the values of the mean and the standard deviation, calculate z.

2. Use the Table of Areas Under the Standard Normal Distribution to determine the proportion of area between two scores.

3. Convert proportions of areas under the normal curve into percentile ranks.

4. To get rid of negative z-scores, know how to make and interpret a T-score transformation.

CHAPTER REVIEW

As we saw in the chapter, the standard deviation provides a powerful tool for making precise quantitative statements about values of a variable when the distribution of a variable approaches the form of the normal curve. We say "approaches" because the normal curve is a theoretical abstraction. The curve never touches the horizontal axis. Mathematically, we say it is asymptotic to the baseline. No matter how deviant the value of a variable is, there will always be some area of the curve beyond that value. In contrast, empirical distributions do have points beyond which no values are found. For example, there are no humans with negative height or negative weight. For this reason, the ideal form of the normal curve is never found in empirical distributions in the behavioral sciences. However, some empirical curves sufficiently approach the form of the theoretical normal curve that we can validly and advantageously apply the properties of the normal distribution to these empirical distributions.

There are a number of helpful calculations we can make if we know the mean and standard deviation of a data set that is drawn from a population that approaches normality.

First of all, we can ascertain how far a raw score is from the mean in terms of standard deviation units. The process of dividing the deviation of a score from the mean by the standard deviation is known as the transformation to standard scores or the *z-score-transformation*. To illustrate, the deviation of a score of 59 from a mean of 52 is 7. If the standard deviation is 3.5, the *z*-score equals 7/3.5 = 2. In other words, the raw score of 59 is two standard deviation units above the mean. In equation form:

$$z = \frac{X - \bar{X}}{s}$$

As you might have observed, a raw score that is less than the mean yields a negative *z*-score. On the other hand, a score that is greater than the mean yields a positive *z*-score. Finally, a score that equals the mean results in a *z*-score that is equal to zero. For practice, find the *z*-scores of the following values of a variable—10, 12, 21—when $\bar{X} = 12$ and $s = 5$. The answers are, respectively, –.40, 0, and 1.80.

One of the advantages of transforming raw scores to *z*-scores for normally distributed variables is that it enables us to compare an individual's performance on two or more unrelated tasks. Suppose we wanted to compare an astronaut's score on a manual dexterity test with her comprehension of an abstract verbal task. If we knew only that she scored 20 on the dexterity test and 540 on the verbal task, we could say little about her relative performance on each task. But suppose we knew that the mean and standard deviation of scores on the dexterity test is 18 and 2, respectively. Her *z*-score on manual dexterity would be *z* = (20 – 18)/2 = 1.00. Thus, relative to the mean, her score is one standard deviation above the mean. Suppose next that the mean and standard deviation on the abstract verbal task is 500 and 20, respectively. Her *z*-score would be (540 – 500)/20 = 2.00. Since both *z*-scores are positive, we know that she scored higher than the mean on each task. Moreover, since her *z*-score on the abstract verbal task (*z* = 2.00) is higher than her *z*-score on the dexterity test (*z* = 1.00), we know that she performed higher on the verbal task than she did on the dexterity task.

If these tasks are normally distributed, the standard normal distribution may be used to characterize more precisely each *z*-score. When plotted on a graph, any normal distribution is symmetrical and bell-shaped. In the standard normal distribution, the mean is equal to zero and the standard deviation equals 1.00. Moreover, the total area under the curve is 1.00, making it easy to express areas above and below various *z*-scores as proportions of total area. By multiplying these proportions by 100, we can express these areas in terms of percentages of the total area under the curve.

Carefully study the following benchmark properties of the standard normal distribution. They will be useful to you in problem solving throughout this course and beyond.

1. Between the mean and one standard deviation above the mean there is located 34.13% of all cases (area) in a normal distribution. This same percentage is found between the mean and 1 standard deviation *below* the mean. Therefore, we can say that 34.13 + 34.13 = 68.26% of all cases (area) fall between the *z*-scores of –1 and +1.

2. Similarly, between the mean and two standard deviations above the mean there lies 47.72% of all cases (area). If we go to the point 2 standard deviations below the mean and consider the percentage of cases between this point and the mean, we'll find the identical percentage of cases—47.72%. Thus, 47.72 + 47.72 = 95.44% of cases (area) fall between the *z*-scores of –2 and +2.

3. Let's go to points 3 standard deviations above and below the mean. Between the mean and +3 standard deviations there is 49.87% of the cases.

Since the normal curve is symmetrical, we know that 49.87% of the cases (area) lies between the mean and –3 standard deviations. Thus, 49.87 + 49.87 = 99.74% of the cases (area) falls between the z-scores of –3 and +3. In other words, virtually all of the area falls between ±3 standard deviations of the mean.

Let's look at a few problems in which we apply these features of the standard normal curve.

• Since you know that 50% of the cases in a normal distribution is above the mean and 49.87% falls between the mean and a point 3 standard deviations above it, the percentage of cases lying beyond 3 standard deviations is 0.13. In other words, a z-score of ±3.00 is exceedingly rare. In point of fact, 99.87% of all cases or of the total area are higher in value than a score 3 standard deviations below the mean.

• Between a z-score of 0 and –1 there is 34.13% of the cases. Moreover, 47.72% of all cases is found between 0 and a z-score of –2. Therefore, 47.72 – 34.13 = 13.59% of all cases is found between the z-scores of –1 and –2. The same percentage of cases is also found between z-scores of +1 and +2.

Since the percentile rank of a score represents the percent of cases in a comparison group falling below that score, we can determine the percentile rank of a z-score of –1. We know that 50% of the cases falls below the mean. Of these, 34.13% is found between 0 and –1. Therefore, 50 – 34.13 = 15.87, the percentile rank of a z-score of –1.

Our astronaut obtained a $z = 2.00$ on abstract verbal tasks. Her percentile rank is, therefore, .50 + .4772 or .9772, in decimal form. In other words, she scored better than 97% of the population on which this test was standardized and less than 3% (2.27%, to be precise) scored higher than she. On the manual dexterity task, on which she obtained a $z = 1.00$, her percentile rank is .50 + .3413. Even on this task, she scored higher than 84% of the standardized population.

In actual practice, you will often need to determine the percentages of area (cases) falling between z-scores that are not whole numbers. For this, we must make use of Table A in the appendix of the text. To illustrate, suppose you wished to find the area between the z-scores of 1.4 and 2.4. You first locate the two z-scores in question under column A. You should write down the values under column B corresponding to these two z-scores. The score 1.5 has an area between it and the mean of .4332. Similarly, .4918 is the proportion of all cases between the mean and a z-score of 2.4. Now that you know the areas between the mean and each of the two z-scores, you need only subtract the area corresponding to the smaller z from the area corresponding to the larger z. The same percentage of cases is also found between z-scores of +1 and +2. In the present example, this is .4918 – .4332 = 0.0586. Thus, the proportion of area between $z = 1.4$ and $z = 2.4$ is 0.0586. Expressed as a percentage of cases (area), it is 5.86%.

Let's quickly review what we know about Table A in the appendix of the text. To locate z-scores, we consult column A. To find the area between the mean and any specific value of z, we consult column B. What is the purpose of column C? It is used to determine the proportion of cases (area) beyond a given z-score. If you were asked to determine the proportion of cases beyond a z of –.92, you would consult column C. First, locate the absolute value of $z = –.92$ (recall that this is equal to .92) under column A. The corresponding value in column C is .1788. Thus, the proportion of area beyond (or below) $z = –.92$ is .1788 and the percent of area is 17.88. Had the z been positive, the percentage of area beyond (or above) $z = .92$ would be 17.88.

In working with z-scores and areas under the curve, we need to keep in mind the position that is relative to the mean of each of the z-scores in a given problem or exercise. If you are looking for the area between two scores on the same side of the mean (i.e., both scores are either negative or positive), you would subtract the smaller area found in column B from the larger area. Finding the area between $z = 1.5$ and $z = .60$ would require the subtraction of the area between the mean and $z = .60$ (.2257) from the area between the mean and $z = 1.5$ (.4332). Thus, $.4332 - .2257 = .2075$.

However, if the two z-scores are on the *opposite* side of the mean, you would *add together* the appropriate proportions of area that are shown under column B. Determining the proportion or percentage of cases lying between -2.34 and .51 would require the addition of their column B values.

Very closely related to the z-score are the T-scores (T for transformed). Just as the z-scores are used to compare seemingly dissimilar measures and to find areas under a standard normal curve, so may T-scores also be used. The primary distinction between z-scores and T-scores is that the transformed scores eliminate negative values and, in most cases, decimal values. These features have been accomplished by moving the z-score distribution to the right along a graphic scale (by adding a constant) and by stretching out the set of scores (by multiplying z by a constant). So, with T-scores, the mean may be 100 and the standard deviation 10. With z-scores, the comparable values are 0 and 1. The formula for converting a z-score to a T-score is

$$T = 100 + 10z$$

As you can see, we are both multiplying by and adding a constant to the original z-score. The constant 100 corresponds to the value of the mean in the T-score distribution, whereas the 10 is equivalent to the standard deviation. Other constants are also used in the T-score transformations. You may already be familiar with T-score transformations in which a constant of 500 is added and the z-score is multiplied by 100—namely, the college entrance exams and the SATs.

If you wish to convert T-scores to the standard normal distribution, you should use the formula.

$$z = (T - \bar{T})/10$$

Once you have converted T-scores to z-scores, you may then determine percentile rankings and proportions of areas that are associated with scores by consulting Table A in the appendix of your text and proceeding as you did previously with z-scores. Thus, a person achieving a score of 620 on one scale of the SAT would have a corresponding z-score of $(620 - 500)/100 = 1.20$. This corresponds to a percentile rank of $.5000 + .3849 = .8849$. Thus, a score of 620 is higher than the score achieved by more than 88% of the standardization group.

SIMPLIFYING THE CALCULATION OF AREAS BETWEEN STANDARD SCORES

You can simplify the calculations of areas between given z-scores by converting each z-score to a percentile rank and subtracting the smaller value from the larger one. To illustrate, in the Chapter Review, we wished to learn the percentage of cases between $z = -2.34$ and $z = .51$. When multiplied by 100, the percentile rank of a negative value of z is given in column C (area beyond z). Consulting column C of Table A under 2.34, we find .0094. When multiplied by 100, the percentile rank is .94. The percentile rank of $z = .51$ is $50 + 100 \times .1950 = 69.50$. The difference in area between the two z-scores then becomes $69.50 - .94 = 68.56$.

SELECTED EXERCISES

1. In the exercises for Chapter 6, we pointed out that the differences between the mean and the scores at the 84th and 16th percentiles often provide a good approximation to the standard deviation of the sample for distributions that are approximately normally distributed. Explain why this is so.

2. Calculate the standard deviation in each of the following distributions. Find the difference between the mean and the score at the 16th percentile. Find the disparity between these approximations and the obtained standard deviations. Does the amount of disparity provide a gross estimate of how well each distribution approximates the normal curve?

X	f_A	f_B	f_C
12	1	2	35
11	4	10	32
10	9	15	20
9	13	28	25
8	25	35	10
7	33	29	6
6	25	12	3
5	13	4	2
4	9	0	1
3	4	1	2
2	1	1	1

3. Explain what is meant by the statement that z-scores represent abstract numbers as opposed to the concrete values of the original numbers.

4. Answer the following:

   a. Between the mean and 1σ below it, the proportion of cases or area is _____ .

   b. The corresponding percentage of area is _____ .

   c. Out of 500 cases, this would consist of _____ below $z = -1.00$ and _____ above $z = -1.00$.

   d. Out of 5000 cases, this would correspond to _____ below $z = -1.00$, _____ above $z = -1.00$, and _____ between $z = -1.00$ and the mean.

5. If two normal distributions have the same mean, are they identical? Why?

6. If the mean on a test was 76 and the standard deviation was 7, what grade was achieved by a person with a z-score of -1.8?

7. If test scores for 80 students were normally distributed with a mean of 76 and a standard deviation of 8, how many students scored higher than 70? How many students scored between 80 and 90?

8. In the table on the right, fill in the percentile ranks of the z-scores.

Percentile Ranks for Various z-scores

z-score	Percentile rank
+3.00	
+2.00	
+1.00	
0	
−1.00	
−2.00	
−3.00	

Answers:

1. In the normal curve, a score 1 standard deviation below the mean corresponds to the point that cuts off the bottom 16% of the distribution. This is, by definition, the 16th percentile. Similarly, a score 1 standard deviation above the mean corresponds to the value that cut off the upper 16%. Thus, it is at the $100 - 16 = 84$th percentile.

2.

Distribution	$\bar{X}$	s	$\bar{X} - X_{16}$
a	7	1.90	1.89
b	8.15	1.67	1.45
c	9.92	2.07	1.73

3. The z-scores represent the deviations from the mean of the values of any variable expressed in terms of standard deviation units.

4. a) 0.3413   b) 34.13%   c) 79.35, 420.65   d) 793.5, 4206.5, 1706.5

5. It depends on the standard deviations.

6. Since $z = (X - \bar{X})/s$, $X = sz + \bar{X}$;   substituting gives $X = (7)(-1.8) + 76 = 63.4$.

7. a) $z_{70} = (70 - 76)/8 = -.75$. Proportion of area above $z = -.75 = .2734 + .50 = .7734$. Thus, number of students above $z = -.75 = 80 \times .7734 = 61.87$, which rounds to 62.

   b) $z_{90} = (90 - 76)/8 = 1.75$; area between the mean and z equals .4599; $z_{80} = (80 - 76)/8 = .5$; area between the mean and z equals .1905; the proportion of difference in areas, $.4599 - .1905 = .2684$; $80 \times .2684 = 21.47$, rounded to 21.

8. Beginning with the percentage corresponding to the z-score, +3.00: 99.87, 97.72, 84.13, 50.00, 15.87, 2.28, 0.13.

SELF-QUIZ: TRUE-FALSE

Circle T or F.

T F   1. A score is meaningful only in relation to some reference group.

T F   2. If $\bar{X} = 10$, $X = 6$, and $s = 3$, then $z = 2.00$.

T F   3. A z-score represents the deviation of a specific score from the mean expressed in standard deviation units.

T F   4. Transforming to z-scores permits us to compare a person's relative position on two different variables.

T F   5. The standard normal distribution has a μ of 1, a σ of 0, and a total area equal to 1.00.

T F   6. A z-score may be defined as X/s only when $\bar{X} = 0$.

T F   7. $X = zs$ when $0 > \bar{X}$ or $\bar{X} > 0$.

T F   8. In a normal distribution, approximately 68% of the area is found between $\pm 1\sigma$.

T F   9. When we transform the scores of a normally distributed variable to z-scores, we are expressing these scores in units of the standard normal curve.

T F   10. If a z of 1.73 includes 45.82% of the area between it and the mean, 4.18% of the area lies beyond that z.

T F   11. If $(X - \bar{X}) = 25$, $\sigma = 16$, and $\mu = 100$, then $z = 0.25$.

T F   12. If the area between a z of $-.75$ and the mean is .2734 and the area between a z of -1.23 and the mean is .3907, the proportion of area between the two scores is .6641.

T F   13. The sum of z-scores is equal to zero.

T F   14. The variance of z-scores is greater than the standard deviation of z-scores.

T F   15. The sum of the squared z-scores is equal to the square root of N.

T F   16. A standard deviation may be regarded as an estimate of precision in measurement.

T F   17. A T-score transformation with a mean of 100 and a standard deviation of 10 would yield no negative values.

T F   18. To convert a T-score into units of the standard normal curve, you should use the formula $z = 100 + 10T$.

T F   19. A standard score differs from a z-score in that it cannot enter into computations.

T F   20. The relative position of z-score with respect to the mean is of no significance in calculating areas under the curve.

Answers: (1) T; (2) F; (3) T; (4) T; (5) F; (6) T; (7) F; (8) T; (9) T; (10) T; (11) F; (12) F; (13) T; (14) F; (15) F; (16) T; (17) T; (18) F; (19) F; (20) F.

SELF-TEST: MULTIPLE-CHOICE

1. The mean of a normal distribution of scores is 100; $s = 10$. The percentage of area between scores 100 and 110 is:

   a) 68   b) 84   c) 50   d) 34   e) 16

2. If the test scores of 400 students are normally distributed with a mean of 100 and σ of 10, the number of students scoring between 90 and 110 is:

   a) 136   b) 272   c) 200   d) 336   e) 68

3. The number of students who scored above 120 in the preceding example is:

   a) 380   b) 9   c) 20   d) 390   e) 2

4. A statistics professor announces that 15% of the grades he gives are A's. The results of the final examination indicate that the mean score is 83, with a standard deviation of 6. What minimum score must a student get to receive an A?

   a) 89   b) 86   c) 95   d) 92   e) 77

5. What percentile does a score represent that is 2.06 standard deviations above the mean in a normal distribution?

   a) 2nd percentile   b) 20th percentile   c) 67th percentile   d) 90th percentile   e) 98th percentile

6. Which z-score corresponds to the 44th percentile?

   a) -1.56 b) $-.44$ c) .15   d) $-.15$ e) 1.56

7. Which z-score corresponds to the 99th percentile?

   a) .99   b) .49   c) 2.33   d) 2.00   e) .01

8. What percentage of the distribution of scores exceeds a z-score of 0?

   a) 0%   b) 25%   c) 50%   d) 100%   e) none of these

9. What proportion of cases in a normally distributed population will have z-scores greater than +1.0 or less than −1.0?

   a) .10   b) .32   c) .50   d) .68   e) 1.0

10. Assume a raw score of 94; the sample mean is 100. The standard deviation is 3 and $N = 88$. The z-score is equal to:

   a) 2.0   b) 3.0   c) 4.0   d) −4.0   e) −2.0

11. To compute a z-score, we need:

   a) a raw score   b) the mean of the distribution   c) the standard deviation of the distribution   d) all of the above   e) none of the above

Questions 12 through 14 refer to a normal distribution with a mean of 100 and a standard deviation of 15.

12. A score of 120 has a percentile rank of:

   a) 9   b) 41   c) 59   d) 82   e) 91

13. The score at the 75th percentile is:

   a) 90   b) 110   c) 123   d) 127   e) between 95 and 105

14. The percentage of people with scores between 75 and 85 is the same as the percentage of people with scores between:

   a) 90 and 100   b) 95 and 105   c) 110 and 120   d) 115 and 125
   e) all of the above

15. Mary obtained a score that was precisely at the mean. In which of the distributions will her z-score be precisely at the mean?

   a) $z = +1.00$ b) $z = 0.00$ c) $z = -3.00$ d) $z = -1.00$
   e) none of the preceding

16. A score of 90 has a percentile rank of:

   a) .67   b) 75   c) 25   d) −.67   e) −25

17. In what kind of distribution will there *not* be a precisely known area 3 standard deviations above *and* below the mean?

   a) normal   b) leptokurtic   c) platykurtic   d) skewed
   e) none of the above

18. Between which values of z is the middle 40% of the area included?

   a) −.25 to .25   b) −.52 to .52   c) −.84 to 84   d) 0 to 1.28
   e) −1.28 to 1.28

19. The total area under the normal curve is:

   a) .999   b) 1.000   c) 1.001   d) 6 standard deviations   e) infinite

20. What is the percentage of area bounded by the interval from 60 to 70 in a normal distribution with a mean of 66 and a standard deviation of 2?

   a) .98   b) 2.15   c) 97.59   d) 97.72   e) a z of 5.00 is not listed

21. You obtain a score of 80 on a test. Which class would you rather be in?

 a) $\bar{X} = 70, s = 10$ b) $\bar{X} = 75, s = 5$ c) $\bar{X} = 60, s = 15$
 d) $\bar{X} = 80, s = 2$ e) $\bar{X} = 76, s = 2$

22. In a normal distribution, what percentage of cases will fall below a score of -1.00?

 a) 16 b) 34 c) 66 d) 84 e) cannot say since more information is required

23. In a normal distribution, what percentage of the cases will fall below a z-score of 1.00?

 a) 16 b) 34 c) 66 d) 84 e) cannot say, since it depends on the form of the distribution

24. In what type of distribution might a score have a percentile rank of 40 and a positive z-score?

 a) normal
 b) leptokurtic
 c) positively skewed
 d) negatively skewed
 e) A score with a percentile rank of 40 will always have a negative z-score.

25. The mean of a distribution of z-scores:

 a) depends on the form of the distribution
 b) depends on the N
 c) depends on the variability
 d) all of the above
 e) none of the above

26. The standard deviation of a distribution of z-scores:

 a) depends on the form of the distribution
 b) depends on the N
 c) depends on the variability
 d) all of the above
 e) none of the above

27. Which of the following is *not* a property of the normal distribution?

 a) Half the raw scores have corresponding negative z-scores.
 b) There are the same number of raw score units between z-scores of .05 and .10 as between z-scores of .0 and .15.
 c) There are raw scores with corresponding z-scores as large as 4.00.
 d) The z-score of the median is 0.00.
 e) All of the above are true statements.

28. The mean of a set of z-scores is:

 a) 1.00 b) 0.00 c) N d) equal to the variance of these z-scores
 e) equal to the standard deviation of these z-scores

29. The variance of a set of z-scores is:

 a) 0.00 b) -1.00 c) 1.00 d) N e) equal to the mean of these z- scores

30. The standard deviation of a set of z-scores is:

 a) 0.00 b) -1.00 c) 1.00 d) N e) equal to the mean of these z-scores

31. A z-score is:

    a) the deviation from the mean divided by the standard deviation
    b) the number of standard deviations a score deviates from the mean
    c) a standard score with a mean of 0.00
    d) all of the above
    e) none of the above

32. Which of the following is *true* for the normal distribution?

    a) Approximately 14% of the cases in the distribution falls between the first and second quartiles.
    b) The 97.5 percentile and $z = 1.96$ have the same score value.
    c) The percentage of cases between plus and minus $z = 3.00$ is equal to exactly 99.0.
    d) A z-score of 1.50 is equivalent to a raw score that is twice that corresponding to a z-score of .75.
    e) If a z-score of 1.75 is equivalent to a raw score of 28, then the raw score -28 has a z-score of -1.75.

33. An industrial firm finds that individuals scoring in the top 1% of the norm group on a specific aptitude test are most likely to be successful executives. Which of the following individuals is most likely to succeed?

    a) Mr. A has a z-score equal to 2.00.
    b) Miss B has a percentile rank of 1.00.
    c) Miss C's score falls between $z = 1.75$ and $z = 1.88$.
    d) Mrs. D's score falls $2\frac{1}{2}$ standard deviations above the mean.
    e) Mr. E's score falls at the 3rd quartile.

Answers: (1) d; (2) b; (3) b; (4) a; (5) e; (6) d; (7) c; (8) c; (9) b; (10) e; (11) d; (12) e; (13) b; (14) d; (15) b; (16) c; (17) d; (18) b; (19) b; (20) c; (21) e; (22) c; (23) d; (24) d; (25) e; (26) e; (27) e; (28) b; (29) c; (30) c; (31) d; (32) b; (33) d.

8

Correlation

BEHAVIORAL OBJECTIVES

Conceptual Objectives

1. Define the functions served by correlation coefficients. State the three factors that influence the decision concerning the appropriate correlation to use.

2. State the characteristics that are common to all correlation coefficients.

3. Define Pearson r in words and in algebraic notation. Describe the relationship between Pearson r and z-scores. Specify when Pearson r is appropriate for describing the degree of relationship between two variables.

4. Identify the formula for the mean deviation method of calculating Pearson r and state when this method is most useful.

5. Identify the formula for the raw score method of calculating Pearson r.

6. State some alternative possibilities for explaining low correlations when Pearson r is used to quantify the extent of a relationship between two variables. Explain how a truncated range might affect the size of the correlation coefficient.

7. Describe when the Spearman r_s is appropriate for ascertaining the correlation between two variables. Define Spearman r_s in words and in algebraic form.

Procedural Objectives

1. When interval/ratio data are provided, calculate a Pearson r using the mean deviation method and the raw score method.

2. Given ordinal data, calculate a Spearman r_s correlation coefficient.

CHAPTER REVIEW

Scientists routinely examine countless hypotheses in which the relationships between and among two or more variables are of crucial importance. Consider the following: What is the relationship, if any, between the incidence of disease and dietary cholesterol, smoking, dietary fiber, vitamins, and regular exercise? Do individuals who exercise regularly show a lower incidence of hypertension and/or heart disease? Is there a relationship between amounts of daily exercise and levels of serum cholesterol? What about those who smoke a pack a day or more? Are they more at risk for respiratory and circulatory disorders? What about child and spouse abusers? Is there a relationship between their abusive behavior and their own childhood experiences with their parents or caregivers?

In raising questions of this sort, we are concerned with the relationship between and among two or more variables. To express quantitatively the extent to which two variables are related (i.e., they vary together or covary), we calculate a statistic known as a *correlation coefficient*. Although there are many types of correlations, we limit ourselves in this chapter to the discussion of two of them—Pearson r and Spearman r (r_s).

Our first correlation coefficient is the Pearson product moment coefficient, or the Pearson r. The Pearson r is used with interval/ratio-scaled data. Although the Pearson r is appropriate for use with a vast percentage of interval/ratio-scaled data, there are instances in which it is not suitable—namely, when the graphic relationship between the two variables does not form a straight line. It is then said to be nonlinear.

Before looking at the Pearson *r* in depth, let's look at a few of the characteristics that hold true for *any* of the correlation coefficients that may be used to quantify the relationship between and among variables.

1. Two sets of measurements are obtained on the same individuals, events, objects, or pairs of individuals who are matched on some basis. In our exercise/cholesterol example, we would obtain two measures on each individual in our study—the amount of daily exercise and the level of serum cholesterol that is found in a sample of blood.

2. The values of the correlation coefficients vary between +1.00 and −1.00. If we were to calculate a correlation coefficient of −1.05 or +3.00, we would know that we had made an error since the maximum possible positive correlation coefficient is +1.00 and the maximum negative is −1.00. Although perfect relationships between variables are common in fields like mathematics (the correlation between the length of one side of a square and the circumference of a square is +1.00), such high correlations are rare in the empirical sciences. This is due to the fact that many factors (including chance variations) influence the values of empirical variables.

3. When the values of two variables rise and fall together, there is a positive relationship between the variables. Figure 8.1 shows a scatter diagram of the percentage of individuals in 15 states who were overweight and the percentage

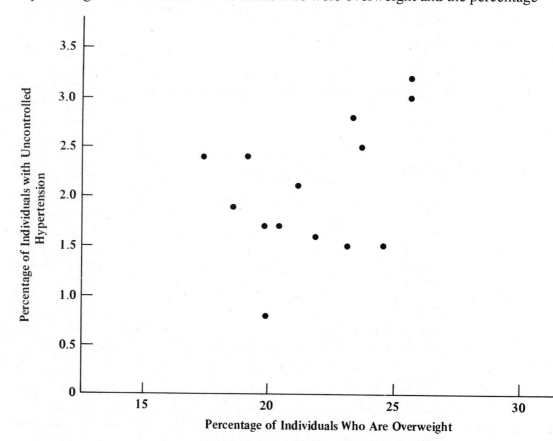

Figure 8.1   **Scatter diagram of percentage of individuals in 15 states who were overweight and the percentage who suffered from uncontrolled hypertension.**

Source: Morbidity and Mortality Weekly Report, 1986, Behavioral Risk Factors, 35, no. 16, 253–254.

who suffered from uncontrolled hypertension. The correlation of $r = 0.41$ shows a moderate positive relationship between overweight and uncontrolled hypertension. It would appear that in states where the percentage of overweight individuals is high there is an increased risk of uncontrolled hypertension.

4. If the values of one variable tends to rise when the values of the second variable fall, there is a negative relationship between the two variables. Some examples of negative correlations include the amount of daily exercise and weight (the more the exercise, the lower the weight); the amount of daily exercise and levels of serum cholesterol (the more the exercise, the lower the levels of serum cholesterol).

One of the requirements for the use of Pearson r as a quantitative measure of the degree of relationship between two variables is that the relationship be linear rather than curvilinear. If the underlying relationship between variables is curvilinear, then the Pearson r will underestimate the extent of the relationship. Figure 8.2 shows two completely determined (mathematical) relationships. In the one on the left dealing with the relationship between the length of one side of a square and the perimeter of a square, the relationship is linear and the Pearson r equals +1.00. In the one on the right (the length of one side and the area of a square), the relationship is curvilinear and the Pearson r is less than +1.00. A curvilinear coefficient (not covered in the text) would show a perfect positive relationship between the length of the side of a square and the area of a square.

When conceptualizing the meaning of Pearson r, you will find it helpful to think in terms of z-scores. Recall from Chapter 7 that a z-score provides information about the relative position of a value of a variable within the distribution. The higher the value of z is, the higher is the corresponding score relative to the rest of the scores in the distribution. Similarly, the higher the absolute value of a negative z is, the lower is the corresponding score relative to the rest of the scores in the distribution. Let's apply these concepts to Pearson r. A high positive correlation means that the individuals, objects, or events being examined occupy the same relative position on the two variables under study. At the opposite pole, a high negative z means that the paired scores are similar in absolute value but opposite in algebraic sign. Thus, if a correlation equals −1.00, a person, object, or event with a $z = 1.86$ on one variable has a corresponding $z = -1.86$ on the second variable. When the correlation equals zero, the z-score

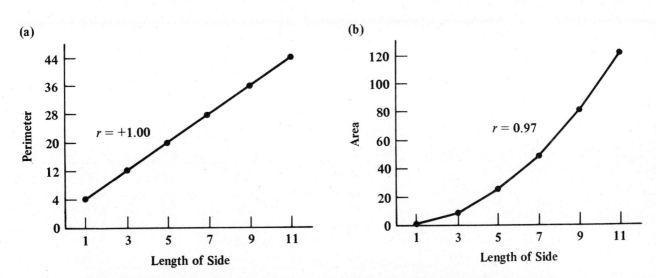

Figure 8.2   Two perfectly positive relationships, one linear (a) and one curvilinear (b). The Pearson r = +1.00 only for the linear relationship.

obtained on one variable bears no relationship to the z-score obtained on the second variable.

There are a number of different computational formulas that may be used to find Pearson r. The following two raw score formulas are straightforward, relatively easily used, and are well-adapted to the pocket calculator and computer applications.

$$r = \frac{\sum XY - \frac{(\sum X)(\sum Y)}{N}}{\sqrt{\left[\sum X^2 - \frac{(\sum X)^2}{N}\right]\left[\sum Y^2 - \frac{(\sum Y)^2}{N}\right]}}$$

$$= \frac{\sum XY - \frac{(\sum X)(\sum Y)}{N}}{\sqrt{SS_X \cdot SS_Y}}$$

The only new term is $\sum XY$. This term instructs you to calculate the sum of the products of X and Y. For each X-value, you multiply by the corresponding Y-value and then add together all the products.

The following are the step-by-step procedures for calculating Pearson r from a data set (Table 8.1) that involves two different measures on each of six different individuals.

Step 1. Count the number of paired scores to find N, N = 6.

Step 2. Find the sum of all the values of X. $\sum X = 375$.

Step 3. Square each value of X and sum the squared values to obtain $\sum X^2$. $\sum X^2 = 26,401$.

Step 4. Repeat steps 2 and 3 for values of Y. $\sum Y = 100$ and $\sum Y^2 = 1822$.

Step 5. Multiply each X by its corresponding Y and find the sum of these products. $\sum XY = 5606$.

Step 6. Find SS_X and SS_Y. $SS_X = 26,401 - (375)^2/6 = 2963.5$, $SS_Y = 1822 - (100)^2/6 = 155.33$.

Step 7. Substitute obtained values in the raw score formula for Pearson r:

$$r = \frac{5606 - (375)(100)/6}{\sqrt{2963.5 \times 155.33}} = \frac{-644}{678.47} = -.95$$

Table 8.1   Scores Made by Six Individuals on Variables X and Y and Various Quantities Required to Calculate Pearson r

Individual	X	X^2	Y	Y^2	X Y
1	90	8,100	10	100	900
2	50	2,500	20	400	1000
3	34	1,156	24	576	816
4	80	6,400	11	121	880
5	82	6,724	15	225	1230
6	39	1,521	20	400	780
Totals N = 6	375	26,401	100	1822	5606

When rounded to two decimal places, Pearson r equals $-.95$. This means that there is a high negative relationship between the two variables. Indeed, it is nearly a perfect negative relationship. Thus, individuals scoring high on variable X scored low on Y, and those scoring low on X scored high on Y.

When one variable is measured on an ordinal scale and the other variable is ordinal or higher (interval or ratio), the Spearman r is the preferred correlation coefficient. For instance, suppose a group of women were ranked according to the level of education that each had attained. Suppose also that they were ranked on a scale of socioeconomic status. If these two ordinal measures were to be correlated, the Spearman r would be the measure of choice. Spearman r is sometimes referred to as the rank correlation coefficient because both variables must be expressed as ranks prior to calculating Spearman r. This is the case even if the scale of one of the variables is interval or ratio.

The formula to use in determining the Spearman r is

$$r_s = 1 - \frac{6\Sigma D^2}{N(N^2 - 1)}$$

In Table 8.2, the X-variable consists of subjective evaluations by a supervisor of eight employees. The higher the numerical value was, the better was the supervisor's evaluation of the performance of the employee. This variable is assumed to be ordinal since the score units cannot be considered equal. The Y-variable is the number of days late to work (20 minutes or more) during a six-month period. Thus, the Y-variable is a discrete ratio variable in which there are two values—late and not late. Rank the X and Y values where necessary and determine the Spearman r.

Table 8.2   Two Values of Two Variables on Each of Eight Employees—Supervisor's Evaluation of Work Performance (X) and Number of Days Late to Work (Y)

Employee	X	Y	Rank of X	Rank of Y	Difference in ranks
1	30	12			
2	24	16			
3	42	10			
4	39	8			
5	45	14			
6	40	7			
7	20	13			
8	28	12			

Note that when there are ties, the rank assigned consists of the mean of the ranks that would have been assigned if there were no ties. Thus, in Table 8.2, subjects 1 and 8 are tied in terms of the number of latenesses. Without ties, they would have been assigned the ranks of 4 and 5. Accordingly, they are each assigned the mean rank of 4 and 5 (9/2) which is 4.5.

Table 8.3 shows the assignment of ranks, the differences and squared differences, and the steps in the calculation of Spearman r.

Step 1.   Assign ranks to each of the two variables, using low numerical ranks for high values and vice versa. (*Note:* If you assigned them in the reverse order, you would obtain the same Spearman r. Just be consistent in what you do.)

Step 2.   Subtract each Y rank from each X rank to obtain the difference. Keep track of the sign of the difference.

Table 8.3   Ranks on Two Variables Shown in Table 8.2 and Calculation of Spearman r

Employee	X	Y	Rank of X	Rank of Y	Difference	D^2
1	30	12	5	4.5	0.5	0.25
2	24	16	7	1	6.0	36.00
3	42	10	2	6	−4.0	16.00
4	39	8	4	7	−3.0	9.00
5	45	14	1	2	−1.0	1.00
6	40	7	3	8	−5.0	25.00
7	20	13	8	3	5.0	25.00
8	28	12	6	4.5	1.5	2.25
			Sum 36	36	0	114.50

Step 3.   Square each difference and then find the sum. In the present example, the sum is 114.50.

Step 4.   As a check on accuracy, obtain the sum of the X, Y, and difference columns. The sum of the X column should equal the sum of the Y column, and the sum of the difference column should equal zero. If the sums of the X and Y columns do not agree or if the sum of the difference column is not zero, you have made either an error in ranking or in arithmetic.

Step 5.   Substitute $\sum D^2$ and N(the number of paired ranks) in the formula for Spearman r. You should obtain

$$r_s = 1 - (6 \times 114.5)8 \times 63$$
$$= 1 - 681/504$$
$$= 1 - 1.36 = -.36$$

As we can see, there appears to be a small negative correlation between supervisory evaluation and lateness performance. In other words, if you were frequently late, you would tend to obtain a poorer evaluation than if you were regularly on time. In inferential statistics, we would raise the question as to whether this sample correlation was sufficiently high to rule it out as a chance occurrence.

Now a word for computer buffs. If you have prepackaged statistical software, you will not often find the Spearman r. Well, it turns out that the Spearman r with ranks is based on exactly the same mathematical foundation as the Pearson r with score data. Thus, if you have a statistical program that computes Pearson r, you can also use that program to calculate Spearman r. Keep in mind that you *must input the ranks* when the paired scores are requested. The computed Spearman r will be identical to the one obtained when the formula for r_s is used except when there are many tied ranks. In the case of many ties, the Pearson r, applied to ranks, yields a more accurate correlation coefficient.

SELECTED EXERCISES

1. Determine the Pearson product moment coefficient of correlation for the following set of data.

Individual	X	X^2	Y	Y^2	XY
A	2	_____	5	_____	_____
B	8	_____	9	_____	_____
C	9	_____	12	_____	_____
D	5	_____	7	_____	_____
E	7	_____	4	_____	_____
F	3	_____	2	_____	_____
G	11	_____	13	_____	_____
H	10	_____	8	_____	_____
I	8	_____	10	_____	_____
J	6	_____	5	_____	_____
$\sum X =$ _____	$\sum X^2 =$ _____		$\sum Y =$ _____	$\sum Y^2 =$ _____	$\sum XY =$ _____

a) $SS_X = \sum X^2 - \dfrac{(\sum X)^2}{N}$ 　　　b) $SS_Y = \sum Y^2 - \dfrac{(\sum Y)^2}{N}$

　　$=$ _____ 　　　　　　　$=$ _____

c) $\sum (X - \bar{X})(Y - \bar{Y}) = \sum XY - \dfrac{(\sum X)(\sum Y)}{N}$ 　　d) $r = \dfrac{\sum (X - \bar{X})(Y - \bar{Y})}{\sqrt{SS_X \cdot SS_Y}}$

　　　　$=$ _____ 　　　　　　　　$=$ _____

2. Reverse the order of all the scores in the Y variable in Selected Exercise 1. Recalculate Pearson r.

3. Unlike the sum of squares, which cannot assume a negative value, the sum of the cross products can be negative. Under what circumstances will the sum of the cross products be negative?

4. A sales manager maintains that his best salespeople possess outstanding leadership qualities. To test his hypothesis, a group of salespeople are independently rated on leadership qualities. Find the correlation between rank in leadership and gross annual sales (hypothetical data).

Sales-people	Rank in leadership	Gross sales in thousands	Rank in gross sales	D	D^2
A	1	203	_____	_____	_____
B	2	196	_____	_____	_____
C	3	207	_____	_____	_____
D	4	180	_____	_____	_____
E	5	135	_____	_____	_____
F	6	157	_____	_____	_____
G	7	178	_____	_____	_____
H	8	193	_____	_____	_____
I	9	140	_____	_____	_____
J	10	120	_____	_____	_____
K	11	136	_____	_____	_____
L	12	115	_____	_____	_____
M	13	98	_____	_____	_____
N	14	115	_____	_____	_____
O	15	112	_____	_____	_____
P	16	116	_____	_____	_____
				$\sum D^2 =$	_____

a) $6 \sum D^2 =$ _____   b) $N(N^2 - 1) =$ _____

c) $r_s = 1 - \dfrac{6 \sum D^2}{N(N^2-1)} =$ _____

5. Determine the Pearson product moment coefficient of correlation for the following data.

Individual	X	X^2	Y	Y^2	XY
A	123	_____	26	_____	_____
B	128	_____	45	_____	_____
C	122	_____	31	_____	_____
D	127	_____	46	_____	_____
E	125	_____	38	_____	_____
F	127	_____	38	_____	_____
G	126	_____	29	_____	_____
H	129	_____	36	_____	_____
I	120	_____	11	_____	_____
J	131	_____	46	_____	_____
K	124	_____	35	_____	_____
$\sum X =$ _____	$\sum X^2 =$ _____		$\sum Y =$ _____	$\sum Y^2 =$ _____	$\sum XY =$ _____

[*Hint:* Subtract 120 from each X score to simplify calculations.]

6. Determine the Pearson product moment coefficient of correlation for the following data.

Individual	X	X^2	Y	Y^2	XY
A	5	_____	24	_____	_____
B	10	_____	20	_____	_____
C	20	_____	16	_____	_____
D	35	_____	16	_____	_____
E	45	_____	12	_____	_____
F	50	_____	20	_____	_____
G	60	_____	8	_____	_____
H	60	_____	4	_____	_____
I	65	_____	12	_____	_____
J	70	_____	4	_____	_____
K	70	_____	8	_____	_____
L	75	_____	4	_____	_____
$\sum X =$ _____	$\sum X^2 =$ _____		$\sum Y =$ _____	$\sum Y^2 =$ _____	$\sum XY =$ _____

7. Divide each X score in Selected Exercise 6 by 5 and each Y score by 4. Recalculate Pearson r.

8. The following table shows seven behavioral risk factors in 15 different states. Calculate the intercorrelations between all possible pairs of comparisons and prepare an intercorrelation matrix.

Behavioral Risk Factors in 15 States, 1984:   Proportion of Individuals at Risk in Seven Behavioral Categories

State	Over-weight	Sedentary life-style	Uncontrolled hypertension	Binge drinking	Heavier drinking	Drinking & driving	Seatbelt nonuse
AZ	20.4	39.5	1.7	20.8	12.0	6.3	61.1
CA	18.6	42.2	1.9	20.4	10.5	4.2	51.2
ID	21.8	46.3	2.1	17.8	5.8	4.2	71.0
IL	23.2	53.8	1.5	22.8	10.2	6.9	68.2
IN	23.7	53.7	2.5	16.6	8.1	4.7	73.3
MN	19.9	49.4	0.8	25.3	7.7	6.9	71.0
MT	19.9	50.0	1.7	27.0	6.9	8.2	70.8
NC	23.4	50.1	2.8	14.2	6.8	4.6	71.3
OH	25.7	52.5	3.0	22.5	8.7	7.4	68.6
RI	19.2	59.9	2.4	19.2	8.6	5.0	71.4
SC	21.9	54.9	1.8	11.0	5.7	2.0	66.6
TN	21.4	60.9	1.6	8.6	4.8	3.3	71.8
UT	17.4	42.7	2.4	10.5	3.2	3.9	66.8
WV	25.7	60.7	3.2	11.6	5.7	2.9	75.9
WI	24.6	50.4	1.5	28.9	10.3	11.3	67.9

Source: Morbidity and Mortality Weekly Report, 1986. Behavioral Risk Factors, 35, no. 16, 253–254.

Answers:

1.

X^2	Y^2	XY
4	25	10
64	81	72
81	144	108
25	49	35
49	16	28
9	4	6
121	169	143
100	64	80
64	100	80
36	25	30
Σ 553	677	592

a) $SS_X = 553 - \dfrac{(69)^2}{10} = 76.9$ b) $SS_Y = 677 - \dfrac{(75)^2}{10} = 114.5$

c) $\Sigma(X - \bar{X})(Y - \bar{Y}) = 592 - \dfrac{(69)(75)}{10} = 74.5$

d) $r = \dfrac{74.5}{\sqrt{(76.9)(114.5)}} = 0.79$

2. $\Sigma(X - \bar{X})(Y - \bar{Y}) = 552 - 517.5 = 34.5$ $r = \dfrac{34.5}{93.83} = 0.37$

3. When the correlation is negative.

4. a) $6\Sigma D^2 = 501.0$

    b) $N(N^2 - 1) = 4080$

    c) $r_s = 1 - \dfrac{501}{4080} = 0.88$

Rank in gross sales	D	D^2
2	−1.0	1.00
3	−1.0	1.00
1	2.0	4.00
5	−1.0	1.00
10	−5.0	25.00
7	−1.0	1.00
6	1.0	1.00
4	4.0	16.00
8	1.0	1.00
11	−1.0	1.00
9	2.0	4.00
13.5	−1.5	2.25
16	−3.0	9.00
13.5	0.5	0.25
15	0.0	0.00
12	4.0	16.00

5.
$$SS_X = 104.55$$
$$SS_Y = 1068.55$$
$$\sum(X - \bar{X})(Y - \bar{Y}) = 274.55$$
$$r = \frac{274.55}{334.24} = 0.82$$

6.
$$SS_X = 6522.92$$
$$SS_Y = 526.67$$
$$\sum(X - \bar{X})(Y - \bar{Y}) = -1588.33$$
$$r = \frac{-1588.33}{1853.49} = -0.86$$

7.
$$SS_X = 260.92$$
$$SS_Y = 32.92$$
$$\sum(X - \bar{X})(Y - \bar{Y}) = -79.42$$
$$r = \frac{-79.42}{92.68} = -0.86$$

8.

	Over-weight	Sedentary life-style	Uncontrolled hypertension	Binge drinking	Heavier drinking	Drinking & driving	Seatbelt nonuse
Weight	—	.459	.413	.051	.128	.193	.454
Life-style	(15)	(15)	.235	−.292	−.294	−.209	.675
Hypertension	(15)	(15)	(15)	−.436	−.263	−.361	.264
Binge drink	(15)	(15)	(15)	(15)	.691	.898	−.187
Heavier drink	(15)	(15)	(15)	(15)	(15)	.567	−.499
Drink & drive	(15)	(15)	(15)	(15)	(15)	(15)	−.029
Seatbelt nonuse	(15)	(15)	(15)	(15)	(15)	(15)	(15)

Note the high correlation between binge drinking and drinking and driving (.898) and the relatively high correlation between sedentary life-style and seatbelt nonuse. It appears that in states where the percentage of binge drinking is high, the percentage of drinking and driving is also high. Surprisingly, the negative correlation (−.499) between heavier drinking and seatbelt nonuse suggests that conformity to seatbelt use is higher in states where heavier drinking is high. In Chapter 12, we'll look at these intercorrelations from the inferential point of view. Specifically, we will ask "Which correlations, if any, are *not* likely to be due to chance and therefore represent real intercorrelations between variables?"

SELF-QUIZ: TRUE-FALSE

Circle T or F.

T F   1. Correlation is concerned with the relationship between and among variables.

T F   2. A correlation coefficient expresses *quantitatively* the extent of the relationship between two variables.

T F   3. A scatter diagram with an ellipse extending from the lower left-hand to the upper right-hand corner represents a negative correlation.

T F   4. If the points in a scatter diagram form a circle, there is a high negative correlation.

T F   5. In a perfect positive correlation, each individual obtains approximately the same z-score on each variable.

T F   6. To calculate a correlation coefficient, it is essential that both variables be on the same scale of scores.

T F   7. In the event of a perfect positive correlation, $\Sigma(z_X z_Y)$ equals N.

T F   8. When $\Sigma(X - \bar{X})(Y - \bar{Y})$ is zero, the correlation is negative.

T F   9. $\Sigma(X - \bar{X})(Y - \bar{Y}) = \Sigma XY - \dfrac{(\Sigma X)(\Sigma Y)}{N}$

T F   10. Correlation coefficients are likely to be spuriously high when there is a truncated range.

T F   11. To apply the r_s formula, only one scale needs to be expressed as ranks.

T F   12. In the event of tied scores when calculating Spearman r, the mean rank is assigned to the tied scores.

T F   13. If ΣD^2 is negative, the correlation is negative.

T F   14. If the relationship between two variables is curvilinear, the Pearson r might deceptively show little or no correlation.

T F   15. A study of alcoholism in Harlem conducted in the Harlem high schools would involve a truncated range.

T F   16. The Spearman r is best used with interval or ratio-scaled data, whereas the Pearson r is preferred with ranked variables.

T F   17. In a scatter diagram, every point represents two values.

T F   18. In a situation in which two variables have dissimilar ranges, a range of 10 versus a range of 75, the Pearson r should be calculated as the measure of correlation.

T F   19. If you were given a list of subject scores on two variables and instructed to determine the Pearson r, you should use this formula or its equivalent:

$$r = \frac{\Sigma(X - \bar{X})(Y - \bar{Y})}{\sqrt{SS_X \cdot SS_Y}}$$

Answers: (1) T; (2) T; (3) F; (4) F; (5) F; (6) F; (7) T; (8) F; (9) T; (10) F; (11) F; (12) T; (13) F; (14) T; (15) T; (16) F; (17) T; (18) F; (19) T.

SELF-TEST: MULTIPLE-CHOICE

1. If one gets a product moment correlation coefficient of .50, then the rank correlation coefficient will be approximately:

   a) 0   b) .25   c) .50   d) 1.0   e) −.50

2. A correlation between college entrance exam grades and scholastic achievement was found to be −1.08. On the basis of this you would tell the university that:

   a)   the entrance exam is a good predictor of success
   b)   they should hire a new statistician
   c)   the exam is a good test
   d)   students who do best on this exam will make the worst students
   e)   students at this school are underachieving

3. It is possible to compute a coefficient of correlation if one is given:

   a)   a single score
   b)   two sets of measurements on the same individuals
   c)   50 scores of a clerical aptitude test
   d)   all of the above
   e)   none of the above

4. You have correlated speed of different cars with gasoline mileage; $r = .35$. You later discover that all speedometers were set 5 miles per hour too fast. You recompute r, using corrected speed scores. What will the new r be?

   a) −.30   b) −.40   c) −.07   d) .35   e) −.35

5. You have correlated height in feet with weight in ounces; $r = .64$. You decide to recompute, after you have divided all the weights by 16 to change them to pounds What will the new r be?

   a) .04   b) .40   c) .16   d) .48   e) .64

6. Truncated range:

   a) increases r b) decreases r c) does not affect r d) is a concept that does not involve r e) none of the above

7. If z_X does not equal z_Y, r is equal to:

   a) 1.00   b) .00   c) .50   d) any value between .00 and ±1.00
   e) none of the above

8. The correlation between midterm and final grades for 300 students is .620. If 5 points are added to each midterm grade, the new r will be:

   a) .124   b) .570   c) .620   d) .670   e) .744

9. After several studies, Professor Smith concludes that there is a zero correlation between body weight and bad tempers. This means that:

   a)   heavy people tend to have bad tempers
   b)   skinny people tend to have bad tempers
   c)   no one has a bad temper
   d)   everyone has a bad temper
   e)   a person with a bad temper may be heavy or skinny

10. Which of the following statements concerning Pearson r is *not* true?

   a) $r = 0.00$ represents the absence of a relationship
   b)   the relationship between the two variables must be nonlinear
   c) $r = .76$ has the same predictive power as $r = −.76$

d) $r = 1.00$ represents a perfect relationship
e)   all of the above are true statements

11. The correlation between IQ and school performance was found to be .64 for the 8000 students at the State U. When the same study was conducted for the 2000 students in the honors program, the obtained correlation coefficient was only .16. Which of the following might explain the difference between the coefficients?

   a)   there were four times as many students in the first study; thus the correlation was four times as great
   b)   the students in the honors program represent a more homogeneous group; thus, the problem was truncated range
   c)   the students in the university are not as intelligent as those in the honors program
   d)   the relationship between IQ and school performance is curvilinear for the second group
   e)   the statistician made an error

12. For which value of r will the z-scores of the X-variable be identical to the corresponding z-scores on the Y-variable?

   a) $r = -1.00$ b) $r = 0.00$ c) $r = 1.00$ d) $r = .50$ e) none of the above

13. When the relationship between two variables is curvilinear, the Pearson r will be:

   a) 0.00   b) negative   c) positive   d) some value between $-.50$ and $-.20$
   d) Pearson r will not be appropriate

14. Which of the following situations might give rise to a misleading correlation coefficient?

   a)   restricting the range of one of the variables
   b)   the variables are related in a nonlinear fashion
   c)   the relationship between the two variables is curvilinear
   d)   small N
   e)   all of the above

15. The selection of which type of coefficient to employ depends on which of the following factors?

   a)   scale of measurement of each variable
   b)   nature of the underlying distribution
   c)   type of relationship between the two variables
   d)   all of the above
   e)   none of the above

16. Two variables, X and Y, are to be correlated. The mean of the distribution of scores for the X-variable is 17 and the standard deviation is 0. The Pearson r will be:

   a)   meaningless to calculate   b) low but negative   c) low but positive
   d)   1.00   e) -1.00

17. Two variables, X and Y, are to be correlated. Prior to ranking, the mean of the distribution of scores for the X-variable is 17 and the standard deviation is 0. The r_s will be:

   a) 0.00   b) .50   c) $-.50$ d) 1.00   e) -1.00

18. The correlation coefficient obtained from a single pair of measurements is:

   a) 0.00   b) .50   c) 1.00   d) -1.00 e) impossible to calculate

19. The correlation coefficient obtained with two pairs of measurements (assuming no tied scores for each variable) will be:

    a) either 0.00 or 1.00   b) either 0.00 or –1.00   c) either 1.00 or –1.00
    d) either .50 or –.50   e) impossible to calculate

20. Decreasing the range of one of two variables often causes the correlation between these variables to:

    a) decrease   b) increase   c) remain the same   d) vary randomly

21. Individuals who are moderately concerned about doing well often obtain higher test scores than those who are either very low or very high in their motivation. The Pearson r between performance and drive level is most likely in this instance to be:

    a) 0   b) +.20   c) +.90   d) –.84   e) inappropriate

22. In the previous item, which of the assumptions underlying r is not met?

    a) continuity of measurement
    b) nonrestriction of the range of scores
    c) linearity of relationship
    d) normality of distribution in test performance
    e) all of the assumptions have been met

23. Which of the following assumptions is required for interpreting Pearson r?

    a) linearity   b) homoscedasticity   c) normality
    d) all of the above   e) none of the above

24. Which of the following statements is true for the Pearson r but not for the Spearman r?

    a) a positive r means that a person scoring low on one variable is also likely to score low on the second variable
    b) two sets of measurements must be obtained on the same individuals (or events) or on pairs of individuals (or events)
    c) the value of r must be between +1.00 and –1.00
    d) if the correlation coefficient is substantially less than 0.00, a person with a high rating on one measure will probably have a low rating on the other variable
    e) all of the preceding statements apply to both correlation coefficients

25. In a correlation in which $r = 1.0$:

    a) $\Sigma(X - \bar{X})(Y - \bar{Y}) = 0$ b) $\Sigma(z_X z_Y) = N$
    c) a scatter diagram could not be drawn
    d) the mean deviation formula would become

    $$r = \frac{\Sigma(X - \bar{X})(Y - \bar{Y})}{1.0}$$

    e) none of the above

Answers: (1) c; (2) b; (3) b; (4) d; (5) e; (6) b; (7) d; (8) c; (9) e; (10) b; (11) b; (12) c; (13) e; (14) e; (15) d; (16) a; (17) b; (18) e; (19) c; (20) a; (21) e; (22) c; (23) d; (24) e; (25) b.

9

Regression and Prediction

BEHAVIORAL OBJECTIVES

Conceptual Objectives

1. Given Pearson r equals +1, 0, or –1. Specify the best way to predict the value of one variable from knowledge of another variable.

2. State the formula for a straight line and what each symbol represents.

3. Define the regression line and explain what is meant by "best fit." Specify the formulas for obtaining the slope of the regression line of Y on X and of X on Y, using the value of Pearson r and the standard deviations of X and Y.

4. State the formulas for Y' and X', using the value of Pearson r as well as the means and the standard deviations of the X and Y variables. Explain what the second term to the right of the equal sign ($r\frac{s_y}{s_x}(X - \bar{X})$ or $r\frac{s_x}{s_y}(Y-\bar{Y})$) represents and how the magnitude of Pearson r influences the second term and the resulting prediction.

5. Describe the similarities between the regression line and the mean. Explain the procedures for constructing lines of regression. State the relationship between the X and Y regression lines and the magnitude of r.

6. Define the term "residual variance" in words and in algebraic notation. Similarly, define the standard error of estimate in words as well as in algebraic notation. Describe what happens to the standard error of estimate as the magnitude of r increases. Compare how changes in the magnitude of r affect the unexplained and explained variance.

7. Define and distinguish among the variation of scores around the sample mean, the unexplained variance, and the explained variance. State the relationship among the total variation, the explained variation, and the unexplained variation. Explore how the magnitude of r changes the values of the explained and unexplained variation.

8. Algebraically and in words describe the coefficient of determination (r^2). State what the magnitude of the coefficient of determination indicates. Explain how the magnitude of r^2 is affected by changes in r.

9. Describe the relationship between correlation and causation.

10. Distinguish the relationship between the coefficient of determination and the coefficient of nondetermination, between the unexplained variance and the coefficient of nondetermination.

Procedural Objectives

1. Given the values of $\bar{X}$, $\bar{Y}$, s_x, and s_y, specify the best prediction on the Y-variable from different values of X. When the magnitude of r is given, use the formula for Y' to obtain the best prediction.

2. Calculate the standard error of estimate with specified values of s_y, s_x, and r.

3. Compute the proportion of the variation accounted for when *r* is given. With the quantities of total variation and the explained variation, calculate the value of r^2.

CHAPTER REVIEW

Think for a moment of the number of times in a day that we use our past experiences to predict future outcomes. Often our sense of well-being and even our lives depend on the accuracy of our predictions. When we routinely drive the gas guzzler down the crowded highway, we are, in effect, predicting that no one in the oncoming lane will suddenly swerve on our side of the road. When we study hard, long, and diligently in preparation for an exam, we expect to do well. We also expect the professor to cover the material for which we prepared and not something totally alien to the content of the course. When we go to our favorite restaurant or fast food joint, we expect to achieve gustatory satisfaction. When we make a date to meet a favorite person at a given time and place, we predict that the appointment will not be forgotten. The fact that most of the time our predictions are correct attests to both the predictability of human behavior and our ability to assess prior experiences as the basis for anticipating future outcomes. On the few occasions when our expectations are not confirmed, we may feel betrayed, hurt, or angry.

Admittedly, the majority of predictions that we make in everyday life are subjective, informal, and not often accompanied by reams of data and statistical formulas. Nevertheless, the processes of prediction in real life do not differ greatly from the processes involved in statistical prediction. Both involve the assessment of past records and/or experiences to predict the future and both require that we make an effort to assess the dependability of the past as a guide to the future.

Of course, our study of statistical prediction will necessarily involve statistical concepts, including some that will be familiar to you and others that will not. To begin, let's consider the mean as a basis for prediction. If the mean is the only descriptive information we have, it provides our best prediction for any single score. For example, if we know that in New York City the mean maximum daily temperature during the month of January is 32.5°F, our best prediction for the mean maximum temperature next January 15 will be 32.5°.

However, we usually have more information than merely the mean of a distribution. If we have the raw data, we can determine the standard deviation as well as the mean. With both these items of information, we can comment on the position of one score relative to the other scores. As you know, relative position is statistically translatable to *z*-scores for normally distributed variables. But means, standard deviations, and *z*-scores are not usually sufficient to give us accurate powers of prediction. To predict the score on one variable from the score on a second variable, we should know the relationship, or correlation, between the two variables.

Let us consider a situation in which we wish to predict the score on one variable from a known score on another variable. The staff in the Admissions Office of a small college is interested in predicting the quality point average (QPA) of applying students from the knowledge of their high school averages. To do this, the Admissions Office will need a correlation between high school grades and QPA among its past students.

Although perfect correlations between empirical variables are rare, let's assume a correlation of +1.00 between these two variables. With a correlation of +1.00, we can say that a student's relative position (or *z*-score) will be equal to his or her relative position (or *z*-score) on the other variable. Therefore, if Nancy has a *z*-score of 1.5 on her high school average, she will also have a predicted *z*-score of 1.5 on the QPA she achieves in college.

Since z-scores are expressed in standard deviation units, if we know the mean and standard deviation of the QPA, we can readily translate Nancy's predicted z-score into a raw score. If $\bar{Y} = 1.8$ and $s_y = .5$, her QPA equals $1.8 + (1.5)(.5) = 2.55$, which is a score 1.5 standard deviations above the mean. Although this is a simplified case—we have assumed a perfect correlation between the two variables—we have predicted Nancy's performance on one variable from her performance on another. In most cases, prediction is more complicated. There is usually a multitude of factors that influence performance so that prediction is rarely perfect or even close to it. In order to help you clearly understand how prediction between two variables operates, we will look at the equation of a line and illustrate linear regression.

From your algebra background, you may recall that the equation of a straight line is

$$Y = a + bX$$

where a is the Y-intercept (the value at which the line crosses the Y-axis) and b is the slope of the line.

In the case of a perfect correlation, either positive or negative, you can determine the Y-value (Y' or predicted Y) corresponding to any value of X. If $b = -5$, we know the slope of the line with which we are dealing is -5. Thus, for every increase of 1 in the value of the X-variable, there is a decrease of 5 in the value of the Y-variable.

The Y-intercept, represented by the letter a, is the point on a graph at which the line intersects the Y-axis. If the line crosses the Y-axis at 3, $a = 3$. How do you write the equation for a straight line having a slope of -5 and a Y-intercept of 3?

$$\begin{aligned} Y &= 3 + (-5)X \\ &= 3 - 5X \end{aligned}$$

Using the following values of X in the preceding equation of a straight line, solve for Y: $-2, 0, 4$. Answers: $13, 3, -17$.

If the correlation between the two variables is ± 1.00, the scatter diagram for the two variables will form a straight line. The line will have a negative slope if $r = -1.00$ and a positive slope if $r = +1.00$. For every pair of values, X and Y, there will be a point located precisely on the straight line.

But what happens if our correlation is *not* perfect? Our data points will not all lie on a straight line. They will be scattered around the straight line, with the degree of scatter depending on the magnitude of the correlation. If the correlation is very high, most of the points representing paired values of the two variables will lie close to the line. If the correlation is low, the points will be more widely scattered. If $r = 0$, the slope will be zero, the Y-intercept will be the mean of Y, and all points in the line will equal the mean of Y for all values of X. Thus, when $r = 0$, $b = 0$, $a = \bar{Y}$. The straight line becomes: $Y = \bar{Y} + (0)(X)$. The second term to the right in the equation always equals 0, no matter what value of X is substituted. Thus, all values of X lead to the same prediction, namely, the mean of Y. When trying to predict value from one variable to another, knowledge of the linear equation provides absolutely no advantage over knowing only the mean when the correlation between two variables is zero.

When the correlation is other than perfect, there exists no single straight line that contains all the points. What we must find is the straight line that best fits the data. This best-fitting line is known as the regression line.

The regression line, or line of "best fit," is defined as the straight line that makes the squared deviations around it minimal. The deviations referred to are the distances from the points in the scatter diagram to the line of best fit. In a way, the regression line is for two variables what the mean is for one. Recall that the mean is that point in a single-variable data set from which the deviations are zero and squared deviations are minimal. For two-variable problems, the deviations of scores around the regression line are zero and the sum of the squared deviations is minimal. Thus, the sum of deviations taken around any straight line other than the regression line will be greater than zero and the sum of the squared deviations will be greater than the sum of the squared deviations taken around the regression line. However, as we have seen, when $r = 0$, the regression line of Y on X is identical to $\bar{Y}$. Under these circumstances, the sum of the squared deviations around the regression line will equal the sum of the squared deviations around the mean.

When we speak of the line of regression of Y on X, we are referring to the line from which Y-values are predicted from their corresponding X-values. Similarly, the line of regression of X on Y is the line from which X-values are predicted from values of Y. For a visual picture of these two regression lines, turn to Figure 9.2 in the text. As you can see, the points of the scatter diagram are not completely contained on either of the lines. So, although the predictions may be reasonably accurate, they will not be perfect.

Before we examine more closely the deviations of scores from the regression line, let's take a look at finding the slopes of Y on X and X on Y and the Y-intercept for Y on X and the X-intercept for X on Y.

The formula for the slope of Y on X, or the regression line for predicting Y from X, is

$$b_y = r(s_y/s_x)$$

You can see from the formula that you require the correlation between X and Y in addition to the standard deviations of the two distributions before you can calculate the line of regression of Y on X. Given a correlation of $-.42$, $s_x = 21$ and $s_y = 3$, $b_y = (-.42)(3/21) = -.06$.

The formula for the line of regression of X on Y (the line of prediction of values of the X-variable from values of the Y-variable) uses all the same elements: r, s_x, and s_y. However, the positions of the standard deviations are reversed. Thus, $b_x = r(s_x/s_y)$. In the present example, $b_x = (-.42)(21/3) = -2.94$. As you can see, both lines have a negative slope.

The other quantity needed to find the equation of a straight line of Y on X is the Y-intercept value, or a_y. The formula for a_y is

$$a_y = \bar{Y} - b_y\bar{X}$$

For the present illustration, assume $\bar{X} = 30$ and $\bar{Y} = 12$. Then, $a_y = 12 - (-.06)(30) = 13.8$. Thus, the regression line of Y on X crosses the Y axis at a value of 13.8.

To find the point at which the regression line of X on Y crosses the X axis, the following formula is used:

$$a_x = \bar{X} - b_x\bar{Y}$$

For the preceding data, $a_x = 30 - (-2.94)(12) = 65.28$.

To obtain the linear regression equation of the line Y on X, you simply substitute a_y for a, and b_y in the equation of a straight line $a + bX$. Thus, $Y' = a_y + b_yX$. For the present problem, when $X = 0$, $Y' = 13.8 + (-.06)(0) = 13.8$, which is the Y-intercept of the regression line. When $X = 30$ (the mean of X), $Y' = 13.8 + (-.06)(30) = 13.8 - 1.8 = 12$, which is the mean of Y. Note that in linear regression, using the mean of X as the predictor always leads to the mean of Y as the prediction. Similarly, when we are predicting from Y to X, using the mean of Y as the predictor always leads to the mean of X and the prediction. This is another way of saying that the two regression lines intersect each other at their respective means.

Let us briefly review the terminology of prediction. The symbols Y' and X' are usually pronounced "Y prime" or "Y predicted" and "X prime" or "X predicted." Whenever Y *is* the value being predicted from the regression line of Y on X, Y' is the symbol used to represent the predicted value. Likewise, if X is being predicted from the regression line of X on Y, X' is the proper notation for the value being predicted.

The following are the formulas for Y' and X':
When we are predicting Y', the formula reads

$$Y' = \bar{Y} + r\frac{s_y}{s_x}(X - \bar{X})$$

When we are predicting X', the formula reads

$$X' = \bar{X} + r\frac{s_y}{s_x}(Y - \bar{Y})$$

As you can see, the formulas for Y' and X' are very much alike. They were both derived by substituting the correct values of a and b in the equation of a straight line.

To apply these formulas in predicting values of X or Y, you must have specific descriptive information about the two-variable data set, namely, the means, the standard deviations, and the correlation coefficient. To provide practice using the regression line formulas, consider the following hypothetical distributions—expense of extracting a ton of mineral versus the natural concentration of the mineral in pounds per ton of ore.

Natural occurrence (no. of pounds per ton)	Expense per ton
$\bar{X} = 25$	$\bar{Y} = \$60$
$s_x = 10$	$s_y = 20$
$r = -.75$	

As you might expect, the correlation between the concentration of the mineral in ore and the expense of extracting the mineral is negative. Thus, the greater the concentration of the mineral in ore is, the lower is the expense per ton of extracting it. Conversely, the lower the concentration of the mineral is, the greater is the cost per ton of extracting it. Suppose you are a statistical consultant to a mining company and you are given the following concentrations of pounds per ton in samples taken from different locations—40, 10, and 65—and you are asked to predict the expense per ton of extracting the mineral. Record your answers below and then check the worked solutions on page 121.

1. When $X = 40$, $Y' = $.

2. When $X = 10$, $Y' = $.

3. When $X = 52$, $Y' = $.

We previously noted that the regression line is a sort of moving mean and shares some of the characteristics of the mean. Let's expand on the parallel of the mean for univariate distributions and the regression line for two variable (bivariate) distributions. Just as we can calculate a standard deviation from the mean based on the deviations of values of the variable from the mean, so also can we calculate a standard deviation from the regression line based on the deviations of scores from the regression line. This standard deviation is called the standard error of estimate. Just as the standard deviation indicates how widely dispersed the scores are about the mean, the standard error of estimate gives us an idea as to how widely dispersed the scores are about the regression line. There are two standard errors of estimate for bivariate data, one around the regression line of Y on X and the other around the regression line of X on Y.

Answers:

1. $Y' = 60 + (-.75)(20/10)(40 - 25) = 60 - (1.50)(15) = \$37.5$

2. $Y' = 60 + (-1.5)(10 - 25) = 60 + 22.5 = \$82.5^*$

3. $Y' = 60 + (-1.5)(52 - 25) = 60 - 40.5 = \$19.5^*$

Either standard error of estimate could be obtained in much the same way as the mean deviation formula is used to calculate the standard deviation of a distribution. The predicted score (analogous to the mean) could be subtracted from each value of the variable, squared, summed, and divided by the number of paired scores. The square root of this value would yield the standard error of estimate. However, as with the calculation of the standard deviation via the mean deviation method, the procedures are time-consuming and cumbersome. The following is the formula for computing the standard error of estimate around the regression line of Y on X:

$$s_{est_y} = s_y \sqrt{1 - r^2}$$

The comparable formula for the standard error of estimate around the regression line of X on Y is:

$$s_{est_x} = s_x \sqrt{1 - r^2}$$

Using the mineral example, let's find the standard error estimate around the regression lines of Y on X (called the standard error of estimate of Y, s_{est_y}) and the standard error of estimate around the regression line of X on Y (called the standard error of estimate of X, s_{est_x}):

$$s_{est_y} = 20\sqrt{1 - (-.75)^2} = (20)(.6614) = 13.23$$
$$s_{est_x} = (10)(.6614) = 6.61$$

Let's explore the deviations and the sum of the squared deviations in greater depth. There are actually three distinct deviations that we can consider and, consequently, three sums of squares of these deviations.

First, you should be acquainted with the squared deviation of scores about the mean, represented by the notation $(Y - \bar{Y})^2$. For the words "squared deviation," we shall substitute "variation" in the remainder of this discussion. So, $\Sigma(Y - \bar{Y})^2$ represents the variation of scores around the mean. This variation is actually the total variation within the distribution. However, we can divide the total variation into two components, explained and unexplained variation.

Explained variation is the variation of predicted scores around the mean, $(Y' - \bar{Y})^2$. You may not realize that the variation of predicted scores around the mean is, in fact, an underestimate of the total variation, that is, the actual variation around the mean. When you consider that the distance from a predicted score to the mean is less than or equal to, *but never greater than,* the distance from the

---

*Note that $(-.75)(20/10)$ is common to all three solutions. Since $(-.75)(20/10) = -1.5$, the quantity -1.5 is directly substituted for $(-.75)(20/10)$ in Solutions 2 and 3.

actual score to the mean, you will see that $\Sigma(Y'-\bar{Y})^2$ is indeed less or equal to the quantity $(Y'-\bar{Y})^2$. In other words, the variation of predicted scores around the mean is generally less than the variation of actual scores around the mean. Of course, with a perfect correlation, our predictions are 100% accurate. In the case of a +1.00 correlation, the explained variation will equal the total variation.

The second component of total variation is unexplained variation. In effect, unexplained variation is the variation of scores around the regression line, given by the notation $(Y - Y')^2$. All of the variation that is not accounted for in terms of the correlation between the two variables constitutes unexplained variation. If the correlation is perfect, Y will equal Y'. The value of the term $\Sigma(Y - Y')$ is then zero.

With a correlation of +1.00 all of the variation is explained; therefore, none of it is unexplained.

Total variation equals explained variation plus unexplained variation. In equation form this is

$$\Sigma(Y -\bar{Y})^2 = \Sigma(Y'-\bar{Y})^2 + \Sigma(Y - Y')^2$$

The two components of total variation are explained and unexplained variation.

From the three quantities, total, explained, and unexplained variation, we can derive two useful ratios, the coefficient of determination (r^2) and the coefficient of nondetermination (k^2). The formulas are

$$r^2 = \frac{\text{explained variation}}{\text{total variation}} = \frac{\Sigma(Y'-\bar{Y})^2}{\Sigma(Y -\bar{Y})^2}$$

and

$$k^2 = \frac{\text{unexplained variation}}{\text{total variation}} = \frac{\Sigma(Y -Y')^2}{\Sigma(Y -\bar{Y})^2}$$

The closer the coefficient of determination is to 1.0, the greater is the amount of explained variation. Similarly, the larger the coefficient of nondetermination is, the greater is the amount of unexplained variation. Then the coefficient of determination involves explained variation, while the coefficient of nondetermination involves unexplained variation.

SELECTED EXERCISES

1. Given $r = 60$, $\bar{X} = 35$, $s_x = 5$, $\bar{Y} = 120$, $s_y = 15$, answer the following:

   a) Michael W obtained a score of 28 on X. What is Y'?
   b) Rachel D obtained a score of 105 on Y. What is X'?
   c) If a score on Y of 140 is considered satisfactory performance on a job, what cutoff value of X should be employed so that 50% of individuals obtaining that score will perform satisfactorily on the average?
   d) Calculate b_y and b_x for the preceding data. Find the square root of the product of the two regression coefficients. Generalize. Prove rigorously that

$$r = \pm\sqrt{b_x \cdot b_y}$$

2. Using the statistics given in Selected Exercise 1, construct the two regression lines for predicting Y from X and X from Y.

3. Describe the relationship among the various sums of squares that may be calculated with correlated data.

4. What can we say about the magnitude of the correlation in the following examples?

   a) if s_{est_y} is large, relative to s_y?
   b) if s_{est_y} is small, relative to s_y?
   c) if $s_{est_y} = s_y$ and $s_{est_x} = s_x$?
   d) if the explained variation is large, relative to the unexplained variation?
   e) if the explained variation is small, relative to the unexplained variation?
   f) if unexplained variation is equal to total variation?
   g) if explained variation is equal to total variation?

5. Given the following data on two forms of the same test:

$\sum X$	= 69		$\sum Y$	= 75
$\sum X^2$	= 553		$\sum Y^2$	= 667
$(\sum X)^2$	= 4671		$(\sum Y)^2$	= 5625
N	= 10	r = .7	N	= 10

   a) calculate the regression equation for predicting scores on the Y-variable
   b) predict Y for $X = 6.0$
   c) predict Y for $X = 8.0$
   d) determine the standard error of estimate of Y

6. Given the following calculations for test X and test Y:

$\sum X$	= 113		$\sum Y$	= 37
$\sum X^2$	= 1325		$\sum Y^2$	= 147
$(\sum X)^2$	= 12,769		$(\sum Y)^2$	= 1369
N	= 12	r = .8	N	= 12

   a) calculate the regression equation for predicting scores on the X-variable
   b) predict X for $Y = 4$
   c) predict X for $Y = 2$
   d) determine the standard error of estimate of X

7. Given the following data for text X and test Y:

X	Y
26	16
16	10
13	12
8	10
6	7
4	5
4	3
2	8
2	2
1	1

   a) calculate the regression equation for predicting scores on the Y-variable
   b) Cye scored 15 on test X; predict his score on test Y
   c) Helen scored –2 on test X; predict her score on test Y
   d) determine the standard error of estimate of Y
   e) calculate the regression equation for predicting scores on the X-variable
   f) Ernie scored 11 on test Y; predict his score on test X
   g) Iris scored 2 on test Y; predict her score on test X
   h) determine the standard error of estimate X

Answers:

1. a) $Y' = 107.4$ b) $X' = 32$ c) $X = 46.11$

   d) $r = \pm\sqrt{(1.8)(.2)} = .60$

$$r = \pm\sqrt{b_x \cdot b_y} = \pm\sqrt{\left(r\frac{s_x}{s_y}\right)\left(r\frac{s_y}{s_x}\right)} = \pm\sqrt{r^2}$$

2.

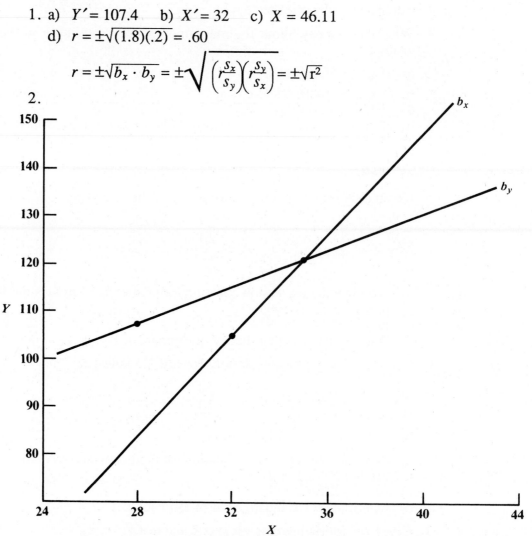

3. SS_x and SS_y reflect the dispersion of the X and Y variables, respectively. The sum of the cross products, $\sum(X - \bar{X})(Y - \bar{Y})$, reflects both the degree and the direction of the relationship between the two variables. If the sum of the cross products is positive, the relationship is positive; if the sum of the cross products is negative, the relationship is negative.

4. a) r is small   b) r is large   c) $r = 0$ d) r is large   e) r is small
   f) $r = 0$ g) $r \pm 1.00$

5. a) $Y' = 7.5 + .77(X - \bar{X})$ b) $Y' = 6.81$ c) $Y' = 8.27$ d) $s_{est_y} = 2.31$

6. a) $X' = 9.42 + 2.25 (Y - \bar{Y})$ b) $X' = 11.49$ c) $X' = 6.99$ d) $s_{est_x} = 3.33$

7. a) $Y' = 3.05 + .53X$ b) $Y' = 11.00$ c) $Y' = 1.99$ d) $s_{est_y} = 2.15$
   e) $X' = -2.68 + 1.47Y$ f) $X' = 13.49$ g) $X' = .26$ h) $s_{est_x} = 3.59$

SELF-QUIZ: TRUE-FALSE

Circle T or F.

T   F   1. If a person scored 2 standard deviations below the mean on one variable, our best guess is that he scored 2 standard deviations below the mean on an unrelated variable.

T F   2. If the correlation between two variables is −1.00, a person obtaining a z of −0.73 on one variable will also obtain a z of −0.73 on the second variable.

T F   3. Monthly salary and yearly income are highly correlated variables.

T F   4. In the formula $Y = a + b_y X$, a and b_y represent constants for a particular set of data.

T F   5. In the formula $Y = a + b_y X$, the letter a represents an atypical score that requires that an adjustment be made in the formula.

T F   6. The regression line may be defined as a straight line that makes the squared deviations around it minimal.

T F   7. The term prediction, as employed in statistics, carries the implication of futurity.

T F   8. The quantity $r \frac{s_y}{s_x}(X - \bar{X})$ represents the predicted deviation from the sample mean resulting from the regression of Y on X.

T F   9. If $r = 0$, $Y' = \bar{Y}$.

T F   10. The two regression lines will have identical slopes only when $r = \pm 1.00$.

T F   11. $z_y' = r z_x$, where $z_y' = Y'$ expressed as a z-score.

T F   12. The regression lines intersect at the means of X and Y only when $r = 0.00$.

T F   13. As r becomes increasingly large, the regression lines rotate away from each other.

T F   14. The regression lines may be regarded as sort of "floating means."

T F   15. The standard error of estimate is a variant of the standard deviation.

T F   16. If the standard error of estimate of Y equals s_y, then $r = 0.00$.

T F   17. When the standard error of estimate of Y equals zero, then $r = \pm 1.00$.

T F   18. The variation $\Sigma(Y - Y')^2$ is known as explained variation.

T F   19. Explained variation is defined in terms of $\Sigma(Y' - \bar{Y})^2$.

T F   20. The total sum of squares consists of two components that may be added together.

T F   21. The coefficient of determination consists of a ratio of explained variation to total variation.

T F   22. The square root of the coefficient of determination equals r.

T F   23. Faulty causal inferences from correlational data are known as the *post hoc* fallacy.

T F   24. The equation for a line with a slope of 10 and a Y-intercept of −12 is $Y = -12 + 10X$.

T F   25. The regression line is the straight line of "best fit."

Answers: (1) F; (2) F; (3) T; (4) T; (5) F; (6) T; (7) F; (8) T; (9) T; (10) T; (11) T; (12) F; (13) F; (14) T; (15) T; (16) T; (17) T; (18) F; (19) T; (20) T; (21) T; (22) T; (23) T; (24) T; (25) T.

SELF-TEST: MULTIPLE-CHOICE

Questions 1 through 9 refer to the following statistics:

$$\bar{X} = 35 \qquad \bar{Y} = 50$$
$$s_x = 5 \qquad s_y = 10$$

1. If $r = 0$, what is the best prediction on the Y-variable for $X = 45$?

   a) 30   b) 40   c) 50   d) 60   e) 70

2. If $r = .50$, what is the best prediction on the Y-variable for $X = 45$?

   a) 30   b) 40   c) 50   d) 60   e) 70

3. If $r = -.50$, what is the best prediction on the Y-variable for $X = 45$?

   a) 30   b) 40   c) 50   d) 60   e) 70

4. If $r = +1.00$, what is the best prediction on the Y-variable for $X = 45$?

   a) 30   b) 40   c) 50   d) 60   e) 70

5. If $r = -1.00$, what is the best prediction on the Y-variable for $X = 45$?

   a) 30   b) 40   c) 50   d) 60   e) 70

6. If $r = .50$, what is the best prediction on the Y-variable for $X = 35$?

   a) 48   b) 49   c) 50   d) 51   e) 52

7. If $r = -.50$, what is the best prediction on the Y-variable for $X = 35$?

   a) 48   b) 49   c) 50   d) 51   e) 52

8. If $r = 1.00$, what is the best prediction on the Y-variable for $X = 35$?

   a) 48   b) 49   c) 50   d) 51   e) 52

9. $s_{est_y} = 0$ when r equals:

   a) 0.00   b) .50   c) –.50   d) 1.00   e) none of the above

10. $s_{est_y} = 0$ when:

    a) $r = 1.00$ b) $r = -1.00$ c) $s_y = 0$
    d) all of the above   e) none of the above

11. $s_{est_y} = s_y$ when:

    a) $r = 0.00$ b) $r = 1.00$ c) $r = -1.00$ d) $s_x = s_y$ e) none of the above

Questions 12 through 14 refer to the following statistics:

$$\bar{X} = 16 \qquad \bar{Y} = 50 \qquad r = .60$$
$$s_x = 2 \qquad s_y = 10$$

12. Approximately 68% of the people who score $X = 16$ will score on the Y-variable:

    a) 50   b) 40–60   c) 42–58   d) 44–56   e) 48–52

13. The best prediction on the Y-variable for $X = 18$ is:

    a) 42   b) 50   c) 56   d) 58   e) 60

14. Approximately 68% of the people who score $X = 18$ will score on the Y-variable?

    a) 40–60   b) 42–58   c) 44–56   d) 48–64   e) 48–68

15. If the correlation between X and Y is 1.00, the angle between the regression lines (plotted in z-scores) is:

    a) 0 degree   b) 45 degrees   c) 90 degrees   d) 180 degrees
    e) none of the above

16. If the correlation between X and Y is 0.00, the angle between the regression lines (plotted in z-scores) is:

    a) 0 degree   b) 45 degrees   c) 90 degrees   d) 180 degrees
    e) none of the above

Questions 17 through 20 refer to the following statistics:

$$\bar{X} = 50 \qquad \bar{Y} = 100 \qquad r = .60$$
$$s_x = 10 \qquad s_y = 5 \qquad N = 1567$$

17. The standard error of estimate of Y is:

    a) 1567   b) 10   c) 8   d) 4   e) 2

18. The standard error of estimate of X is:

    a) 1567   b) 10   c) 8   d) 4   e) 2

19. A predicted Y-score for $X = 50$ will fall between which of the following intervals 68% of the time?

    a) 96–104   b) 98–102   c) 92–108   d) 90–100   e) 95–105

20. A predicted X-score for $Y = 100$ will fall between which of the following intervals 68% of the time?

    a) 48–52   b) 46–54   c) 42–58   d) 40–60   e) 1517–1617

21. The equation $Y = -X$ is an example of:

    a) a negative correlation   b) a straight line   c) neither   d) both
    e) a zero correlation

22. The assumption of linearity in the use of product-moment correlation is made:

    a)   when r is interpreted as a regression line
    b)   when r is interpreted as a correlation coefficient
    c)   for the coefficient of determination
    d)   for the standard error of estimate
    e)   all of the above

23. The assumption of homoscedasticity is necessary in order to:

    a)   write a regression equation
    b)   calculate a correlation coefficient
    c)   correctly interpret the coefficient of determination
    d)   use the standard error of estimate
    e)   all of the above

24. The correlation coefficient is a measure of the slope of the regression line when:

    a) $s_{est_y} = 0$
    b) $r^2 = 1$
    c) scores are in standard score form
    d) scores are in deviate score form
    e) never

25. If the correlation between height and weight is .70, what percentage of the variance in weight would you expect to be associated with the variance in height?

   a) 50   b) 64   c) 49   d) 70   e) not possible to determine

26. If the percentage of variation in one variable associated with another variable is 25, then r must be equal to:

   a) .50   b) .25   c) .90   d) .60   e) .625

27. The standard error of estimate is used to determine:

   a) the most probable score for a given individual on a predicted variable
   b) the reliability of a score actually obtained by an individual on a test
   c) the reliability of a coefficient of correlation
   d) the reliability of a predicted score
   e) the coefficient of determination

28. In predicting a criterion from an aptitude test, one finds that predictions for individuals with low test scores are in actuality about as accurate as in the case of individuals with higher test scores. This finding indicates the presence of:

   a) high correlation   b) high regression   c) high homoscedasticity
   d) normality of distribution   e) all of the above

29. Mr. X took two tests: Test A and Test B. He obtained a score on Test A that was above the mean. (Assume a positive correlation ($r < 1.00$) between tests.) On Test B you would guess that his score would be:

   a) closer to the mean of Test B
   b) further from the mean of Test B
   c) the same distance from the mean of Test B
   d) approximately 1 standard error from the mean of Test B
   e) at the mean of Test B

30. If the correlation between X and Y is 1.00, the angle between the regression line and the X-axis (plotted in z-scores) is:

   a) 90 degrees   b) 0 degree   c) 45 degrees   d) 1.00 degrees
   e) cannot say

31. If the correlation between body weight and annual income were high and positive, we could conclude that:

   a) high incomes cause people to eat more food
   b) low incomes cause people to eat less food
   c) high-income people spend a greater proportion of their income on food
   d) all of the above
   e) none of the above

Answers: (1) c;   (2) d;   (3) b;   (4) e;   (5) a;   (6) c;   (7) c;   (8) c;   (9) d;   (10) d; (11) a; (12) c; (13) c; (14) d; (15) a; (16) c; (17) d; (18) c; (19) a; (20) c; (21) d; (22) e; (23) d; (24) c; (25) c; (26) a; (27) d; (28) c; (29) a; (30) c; (31) e.

PART III

INFERENTIAL STATISTICS: PARAMETRIC TESTS OF SIGNIFICANCE

10

Probability

BEHAVIORAL OBJECTIVES

Conceptual Objectives

1. Know the role of probability theory in inferential statistics. Is probability theory central or peripheral to the making of inferences?

2. Know at what points in an experiment randomness should be introduced. What is a possible undesirable outcome of a failure to achieve randomness?

3. Know the three different approaches to probability, the characteristics of each, and when each approach is appropriate.

4. Know the limits within which the values of probabilities must fall. Show the various ways in which probability may be expressed.

5. Define the addition rule and know under what circumstances it is applied. Know the meaning of the terms mutually exclusive and exhaustive. Know when to apply the addition rule when events are not mutually exclusive and when they are mutually exclusive. Also know how to apply the addition rule when events are both mutually exclusive and exhaustive.

6. Define the multiplication rule. Learn when to apply the multiplication rule to different situations. Distinguish between independent and nonindependent occurrences, paying particular attention to the variation in the multiplication rule formula in each case.

7. Distinguish between sampling with and without replacement. Know the circumstances under which these two methods of sampling may yield large differences in probability values.

8. Distinguish among joint probabilities, marginal probabilities, and conditional probabilities. Know the circumstances under which it is appropriate to use each.

9. Know the formula for obtaining probability of continuous rather than discrete variables. Be able to relate z-scores to probability.

Procedural Objectives

1. Given various sample problems involving discrete variables, calculate probabilities using the addition and multiplication rules.

2. Given frequency data in the form of a 2×2 table, be able to calculate joint, marginal, and conditional probabilities.

3. Be able to construct tree diagrams to calculate the probabilities of conditional events.

4. With the population parameters of continuous variables that are known or estimated, compute one- and two-tailed probabilities of specified events with the aid of Table A in the appendix of the textbook.

CHAPTER REVIEW

With this chapter we are entering a new realm of statistics, that of inferential statistics. All the topics we have covered thus far—percentiles, central tendency, dispersion, standard deviation, correlation and prediction—are categorized under the heading of descriptive statistics. Although the measures of descriptive statistics help us to picture and quantify a data set, conceptually and visually, and to organize and summarize it in convenient and readily usable form, the methods of inferential statistics will enable us to go one invaluable step further. With inferential statistics, we shall learn how to generalize from samples of data to the population from which the samples were taken. For example, if we select a sample of 50 students from the Milwaukee school district to test on certain measures, with techniques of inferential statistics we shall be able to generalize from our sample to the larger population of all Milwaukee schoolchildren.

In this process of generalization, we will use descriptive stastistical measures with which you are already acquainted, such as the mean and standard deviations, as well as inferential statistical methods. We shall begin our section on inferential statistics with a discussion of the basics of probability. You should not take your study of probability lightly, for it includes concepts to which you will be returning time and time again during your involvement with statistics and the behavioral sciences.

Consider these questions: Is 10% of REM (rapid eye movement) sleep out of the total sleep cycle an abnormal amount of REM sleep for an adult? Is the occurrence of male underachievers in school significantly higher than that of female underachievers? When a person is incarcerated for the commission of a crime, what is the probability that the individual will commit another crime after release from imprisonment?

Questions such as these cannot be answered without the aid of probability theory. And, for the behavioral scientist, the ability to answer such questions is essential and exciting. Through the application of inferential statistics, you will learn how to answer these questions.

The first concept to understand is that of randomness. When we are interested in drawing inferences about populations from samples, we want the sample to be representative of the population. One way of achieving representativeness is to use simple random sampling: selecting the sample in such a way that each sample of a given size has precisely the same probability of being selected or, alternatively, selecting the sample in such a way that each member of the population is equally likely to be selected.

Probability theory is concerned with the possible outcomes of experiments. If a scientist is studying weight fluctuation, there are three outcomes for each member in the sample: weight gain, no weight change, and weight loss. To define probability, you should describe it as a theory concerned with the possible outcomes of experiments.

Within probability theory, there are three approaches: classical, empirical, and subjective. In the classical approach, relative frequencies of the possible outcomes are assigned according to theory. The assumption of the classical approach is that an ideal situation exists in which all of the possible outcomes are known or describable. Although we can construct situations in which all of the population members can be counted and their associated probabilities calculated through reasoning rather than experience, it is not always possible to do so. However, let us consider an instance in which all the possible outcomes are known so that we can apply the classical approach to probability.

Consider the toss of a single die, in which there are a limited number of outcomes—six. Assuming that each event has an equal likelihood of occurring, we can calculate the probability of any given outcome by using the probability formula:

$$p(A) = \frac{\text{number of outcomes favoring event } A}{\text{total number of events}}$$

Suppose we are interested in determining the probability that we shall obtain a 5 on the die. Knowing there is only one outcome favoring this event and there are six possible outcomes, we can determine that $p(5) = 1/6$. Calculate the probability that a 7 will appear on the face of the die:

$$p(7) = \frac{0}{6} = 6$$

Since the maximum number that can be obtained using one die is six, the events favoring $p(7)$ are zero.

For most situations in the behavioral sciences, we do not know all possible outcomes or the exact probability that is associated with any single outcome. We must estimate these probabilities on the basis of past observations of samples taken at random from the population. The basic distinction between the empirical and classical approaches is the method of assigning relative frequencies to each of the possible outcomes. With the classical approach, this is accomplished through the assumption of an ideal situation in which the frequencies of the outcomes are known or may be calculated. In the empirical method, relative frequencies are determined from observations. No matter how the probability of single events is obtained, whether by the classical or the empirical approach, the rules for adding and multiplying probabilities remain the same.

Using our example of the die, suppose we are interested in the probability of obtaining an even number (2, 4, or 6) *on a single trial*. To solve this problem, we would need to use the addition rule for mutually exclusive events. The statement of the addition rule is that for mutually exclusive events, the probability of obtaining one or the other of them is equal to the sum of the separate probabilities of each of the events. In symbolic notation,

$$p(2 \text{ or } 4 \text{ or } 6) = p(2) + p(4) + p(6)$$

Since we know that $p(2) = 1/6$ and

$$p(4) = 1/6 \quad \text{and} \quad p(6) = 1/6$$

then the probability of obtaining an even number is:

$$p(2 \text{ or } 4 \text{ or } 6) = 1/6 + 1/6 + 1/6 = 3/6 = 1/2$$

What do you suppose your chances are of obtaining an even number *on two successive tries?* We now must apply the multiplication rule. In the case we have proposed, the multiplication rule could be stated as: The probability of the simultaneous or successive occurrence of two events is the product of the separate probabilities of each event. Restated, the multiplication rule becomes

$$p(A \text{ and } B) = p(A)p(B)$$

in which event A is the occurrence of an even number on the first toss [$p(A) = 1/2$], and event B is the occurrence of an even event on the second toss [$p(B) = 1/2$].

Thus, the probability of obtaining an even number on two successive tries is $p(A \text{ and } B) = 1/2 \times 1/2 = 1/4$.

The partial tree diagram, shown here, illustrates the calculation of the probability of obtaining an even number on two successive tosses of a single die.

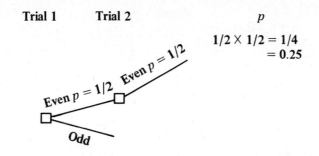

Assume three tosses of a single die. See if you can construct partial tree diagrams showing the following events and their associated probabilities: (a) two even numbers followed by an odd number and (b) three consecutive odd numbers.

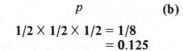

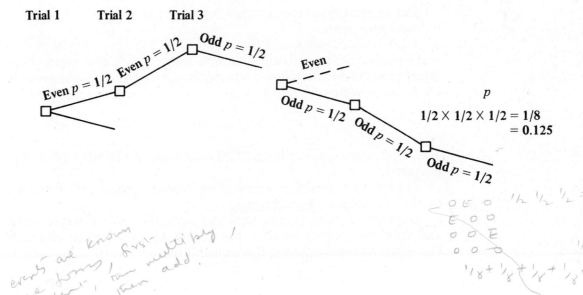

Still assuming three tosses of a single die, construct partial tree diagrams for the following events that combine the multiplication and addition rules: (c) two even numbers out of three tosses; (d) two or more odd numbers out of three tosses.

Until now we have been postulating that our events are independent. Many times, however, psychologists and sociologists find themselves faced with non-independent events. For instance, once a member of a population has been selected for inclusion in a sample, he or she usually is not returned to the population pool. How does this sampling without replacement affect the members yet to be selected for the sample? Rather than being independent, the events are now related or nonindependent.

For example, suppose a person were to select a playing card from a 52-card deck on his or her first trial. On the next trial, there would be only 51 cards remaining. This is an example of sampling without replacement. Again, how does this influence the calculation of probabilities?

(c)

(d)

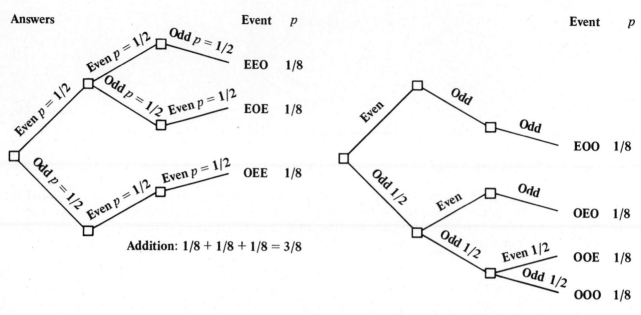

Addition: $1/8 + 1/8 + 1/8 = 3/8$

Addition: $1/8 + 1/8 + 1/8 + 1/8 = 1/2$

Let us restate the multiplication rule, expanding it now to encompass non-independent events:

Given two events A and B, the probability of obtaining both A and B jointly is the product of the probability of obtaining one of these events times the conditional probability of obtaining one event, given that the other event has occurred. Or, in other words,

$$p(A \text{ and } B) = p(A)p(B|A)$$

The conditional probability $p(B|A)$ means the probability of B given the A has occurred.

How do we go about relating the multiplication rule for nonindependent events to problems of probability?

Suppose we wish to determine the probability that a player will select an ace and a king on consecutive tries. Recall that this is sampling without replacement. Our equation for computing this probability is:

$$p(\text{ace and king}) = p(\text{ace})p(\text{king}|\text{ace})$$

Stating the conditional probability $p(\text{king}|\text{ace})$ in words, we would say "the probability that a king will be selected given that an ace has already been selected." The value of $p(\text{ace})$ is 4/52. (Remember that there are four possible aces to be selected.) What happens with the second selection? Assuming that an ace has already been selected, the number of cards remaining on the second attempt is 51. Given that an ace has been selected, the conditional probability that a king will be selected on the second try is 4/51. Substituting the appropriate values, we find that the $p(\text{ace and king}) = 4/52 \times 4/51 = 16/2652 = 0.006$.

Now that we have investigated the addition and multiplication rules of probability, we consider these rules and theory of probability as they relate to continuous variables and the normal curve. For continuous variables, we define probability in a slightly different manner. Instead of regarding the proportion as a ratio of outcomes, we now define it as a ratio of areas in our probability distribution. For continuous variables, the definition of probability is

$$p = \frac{\text{area under portions of a curve}}{\text{total area under the curve}}$$

The curve to which we are referring is based on the standard normal curve model. As you may have guessed, this will enable us to use z-scores and the areas under the curve related to them. Recall that the total area under the normal curve is 1.00. The denominator in the probability proportion then will always be 1.00 when we are dealing with the normal curve.

Using the areas under the normal curve in combination with the addition and multiplication rules of probability theory, we can determine probabilities that are associated with a score occurring singly or jointly with other scores.

Given $\mu = 25$ and $\sigma = 8$, what is the probability of obtaining a score of 31 or less?

$$z = \frac{31 - 25}{8} = .75$$

Checking in Table A in the appendix of the text you will find that 0.2266 is the area beyond a z of 0.75. To find the area under the portion of the curve below a z-score of 0.75, merely subtract 0.2266 from 1.00, the total area under the curve. You will obtain an area of 0.7734. When you set up your probability proportion, you will have

$$p = \frac{.7734}{1.00} = .7734$$

So you see the probability of scoring 31 or lower is .7734.

What is the probability of achieving a score at or below 31 *or* a score of 41 or greater? This sort of question calls for the use of the addition rule of probabilities. When you are requested to find the probability of *either* of two or more events occurring, you should determine the probabilities of the single events and then apply the addition rule. Thus, $z_{\leq 31} = .75$, $p_{\leq 31} = .7734$; $z_{\geq 41} = 16/8 = 2$, $p_{\geq 41} = .0228$.

Therefore, the probability of scoring equal to or less than 31 or equal to or greater than 41 is .7734 + .0228 = .7962.

Frequently, behavioral scientists are interested in the extreme values of a distribution, rather than in the scores clustered about the mean. Sometimes deviant behavior can offer more insight into a problem than can so-called "normal" behavior. Moreover, our statistical decisions are based on our ability to identify regions in a distribution that are extreme, rare, or unusual. To find whether a score on a measure is "extreme," "deviant," "unusual," and so on, a researcher could consult the probabilities that are associated with values at either or both tails of a distribution. If he or she is interested in values at only one tail of the distribution, the proportion is termed a one-tailed p-value. On the other hand, if the scientist is concerned with both ends of the curve, the proportion is called a two-tailed p-value.

A developmental psychologist concerned with children scoring below 60 on a gross motor measure would use a one-tailed p-value. However, an economist wishing to study individuals scoring more than two standard deviations from the mean on a measure of the amount of rent paid per month would have a probability expressed as a two-tailed p-value. One- and two-tailed p-values are topics to which we shall be returning frequently during our study of inferential statistics.

SELECTED EXERCISES

1. What is random sampling? Give two examples.

2. What is biased sampling? Give two examples.

3. When are two events mutually exclusive?

4. When are two events mutually exclusive and exhaustive?

5. When are events independent?

6. Why are mutually exclusive events never independent?

7. A man opens a candy store and a barber shop. In any one year the probability of a robbery in the candy store is .10, and the probability of a robbery in the barbershop is .02. For any one year, what is the probability that:

   a) neither business will be robbed?
   b) both will be robbed?
   c) either one or the other (but not both) will be robbed?

8. Three people work independently at figuring out a puzzle. For each person the probability of success is .25.

   a) what is the probability that the puzzle will be solved?
   b) what is the probability that it will be solved if 8 people work independently on the puzzle (assuming $p = .25$ per person).

9. A manufacturer finds that the cost of manufacturing a certain article is normally distributed with a mean of 48 cents and a standard deviation of 3 cents. What is the probability that his cost will be:

   a) less than 40 cents?
   b) greater than 50 cents?
   c) between 45 and 47 cents?
   d) between 45 and 50 cents?
   e) either less than 45 cents or greater than 50 cents?

10. The results of a magazine survey indicate that the average age of the readers is 37 with a standard deviation of 4. Assuming that these ages are normally distributed, what is the probability of selecting two readers, at random:

    a) who are over 40?
    b) who are between 30 and 35?
    c) one who is over 40, the other who is between 30 and 35?
    d) one who is over 45, the other who is under 35?
    e) who are both over 45, or both under 35?

11. Is it true that when a single score is randomly drawn from a normal distribution, it is practically certain that the score will lie within 3 standard deviations of the mean? Explain.

12. Explain the phrase "as unusual as."

13. Given that $\mu = 92$ and $\sigma = 14$.

    a) what is the probability of obtaining a score of 134 or higher? Is this a one- or two-tailed p-value?
    b) what is the probability of obtaining a score as deviant as 134? Is this a one- or two-tailed p-value?

Answers:

1. Selecting the members of a sample in such a way that, over the long run, each member of the sample has a probability of being selected that equals its proportional representation within the population. Alternatively, selecting samples in such a way that each sample, of a given size, is equally likely to be selected.

2. Selecting the samples or the members of a sample so that certain members are more likely to be selected than their proportional representation in the population and others are less likely to be selected. Telephone polls of the general population are likely to be biased: the very poor, many of whom have no phones, are excluded from the survey. Any "survey" that requires voluntary participation of the respondent such as: write or call and tell us if you favor gun control, abortion rights of women, stronger child support laws, and so forth.

3. When they cannot occur at the same time.

4. When they cannot occur at the same time and, together, they exhaust all possible outcomes.

5. When the probability of the occurrence of one bears no relationship to the occurrence of the other.

6 Mutually exclusive events are related but in a negative fashion: if one occurs, the other cannot.

7. a) $p = (.90)(.98) = .88$ b) $p = (.10)(.02) = .002$
   c)

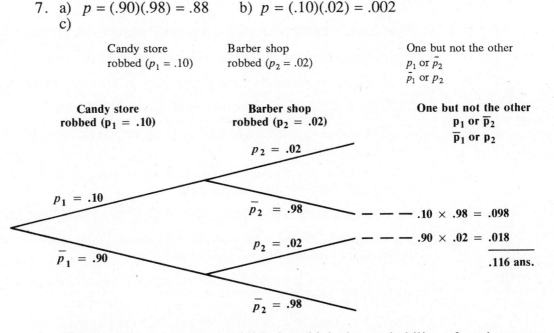

| Candy store robbed ($p_1 = .10$) | Barber shop robbed ($p_2 = .02$) | One but not the other p_1 or $\bar{p}_2$ $\bar{p}_1$ or p_2 |

8. This is a two-category variable in which the probability of each person solving the puzzle is .25 and not solving it is .75. To find the probability that at least one will solve the puzzle, you find the probability that all fail to solve $(.75)^8$ and subtract this from 1.00. Thus, $(.75)^8 = 0.10$ (the probability that none will solve). Therefore, the probability that at least one will solve is $1.00 - .10 = .90$.

9. a) $z = -2.67, \text{p} = .0038$ b) $z = .67, p = .2514$
   c) $z_{45} = -1.00, z_{47} = -.33$, area between $= .3413 - .1293 = .2120$
   d) $z_{45} = -1.00, z_{50} = .67$, area between $= .3413 + .2486 = .5899$
   e) area beyond both $= 1 - .5899 = .4101$

10. a) $z = .75$, area beyond $= .2266, p = .2266^2 = .0513$
    b) $z_{30} = -1.75, z_{35} = -.5$, area between $= .4599 - .1915 = .2684$,
        $p = .2684^2 = .0720$
    c) $p_{>40} = .2266; p_{30-35} = .2684$, joint probability $= (.2266)(.2684) = .0608$
    d) $z_{45} = 2.00, p_{>45} = .0228, z_{<35} = -.5, p_{>35} = .3085$, joint probability $= .0070$
    e) both over 45 $= .0228^2 = .0005$, both under 35 $= .0952$, either both over 45 or under 35 $= .0005 + 0952 = .0957$.

11. The area beyond a z of -3 = .0013; the area beyond a z of 3 = .0013; the probability of selecting a single value beyond the absolute value of $z = 3$ is .0016, or 16 in 10,000. It is almost certain but it will occur 16 times in 10,000, on the average.

12. The expression implies a two-tailed test. For example, obtaining 10 heads on 10 tosses of a single coin is as unusual as obtaining 10 tails.

13. a) $z = (134 - 92)/14 = 3.00$; the area beyond is .0013
     b) a value as rare, unusual, or deviant as 134 includes a z-score as low or lower than -3.00; the two-tailed p-value = .0013 + .0013 = .0026

SELF-QUIZ: TRUE-FALSE

Circle T or F.

T   F   1. If A and B are two mutually exclusive and exhaustive categories, then $p(A) + p(B) = 1.00$.

T   F   2. Mutually exclusive events are never related.

T   F   3. When events are dichotomous and mutually exclusive, $p(A|B) = 0$.

T   F   4. When events are mutually exclusive, $p(A$ and $B) = 0$.

T   F   5. When events are mutually exclusive, $p(A) + p(B)$ is always equal to 1.00.

T   F   6. Collections of random events take on unpredictable forms.

T   F   7. For continuous variables, probability is expressed in terms of proportion of area under a curve.

T   F   8. In symmetrical distributions, two-tailed values may be obtained by dividing the one-tailed probability value in half.

T   F   9. The events "less than three jacks" and "at least three jacks" are independent.

T   F   10. Any observed difference between sample means is the result of unsystematic factors that may vary from study to study.

T   F   11. Events in a series are said to be independent if one event has no predictable effect on the next.

T   F   12. We may accurately generalize our results to the general population from a biased sample.

T   F   13. If we toss 500 unbiased coins, we can accurately predict the proportion that will land "heads" over a lengthy series of tosses.

T   F   14. In the classical definition of probability, the probability of an event is interpreted as an idealized relative frequency of the event.

T   F   15. If the probability of an event occurring equals .06, that event is almost certain to occur.

T   F   16. If $p(A|B) = 1.00$ and $p(B) = 1.00$, the event A is certain to occur.

T   F   17. In a well-shuffled deck of 52 cards, the probability of drawing either a picture card (including the ace) or a spade is 0.48.

T   F   18. When we are sampling with replacement, the events are mutually exclusive.

T   F   19. Employing sampling without replacement, when the population is large relative to the sample, endangers the basic assumptions of probability theory.

T   F   20. In skewed distributions, two-tailed *p*-values may be obtained by doubling the one-tailed *p*-value.

Answers: (1) T; (2) F; (3) T; (4) T; (5) F; (6) T; (7) T; (8) F; (9) F; (10) F; (11) T; (12) F; (13) T; (14) T; (15) F; (16) T; (17) T; (18) F; (19) F; (20) F.

SELF-TEST: MULTIPLE-CHOICE

1. An investigator reports: The mean of the experimental group was five points higher than the mean of the control group. He may conclude that:

   a) the experimental variable had an effect
   b) the experimental variable had only a small effect
   c) the control variable had no effect
   d) the control variable operated to reduce the scores of the control subjects
   e) there is insufficient information to form a conclusion

2. If any event in a series has no predictable effect on another, the events may be said to be:

   a) independent   b) correlated   c) biased   d) reliable   e) systematic

3. If our selections of samples operate to favor certain events over other events, the sample may be said to be:

   a) random   b) biased   c) unsystematic   d) independent
   e) none of the above

4. Distributions of sample statistics based on random sampling from a population:

   a) generally take unpredictable forms
   b) are biased
   c) evidence no reliable relationship with the population parameters
   d) duplicate the parent distribution of scores
   e) generally take predictable forms

5. The question, "What is the probability that four students drawn at random from the student body will have blue eyes?" involves:

   a) probabilities that cannot be ascertained
   b) the classical approach to probability theory
   c) the empirical approach to probability theory
   d) probabilities that are less than 0.00
   e) none of the above

6. If the probability that an event will occur is .10, the odds against this event occurring are:

   a) 10 to 1   b) 1 to 10   c) 9 to 1   d) 1 to 9   e) 19 to 1

7. If the odds in favor of an event occurring are 7 to 1, the probability of this event occurring is:

   a) 1/7   b) 1/6   c) 6/7   d) 5/6   e) 7/8

8. If one is selecting from a well-shuffled 52-card deck of playing cards, the probability of obtaining a 5 is:

   a) 5/52   b) 1/12   c) 1/13   d) 5/47   e) none of the above

9. The odds against drawing a heart from a well-shuffled 52-card deck of playing cards are:

   a) 13 to 1   b) 4 to 1   c) 12 to 1   d) 3 to 1   e) none of the above

10. The probability of selecting a king or queen from a well-shuffled 52-card deck of playing cards is:

    a) 1/13   b) 2/11   c) 2/13   d) 1/12   e) 1/169

11. The probability of selecting a face card (jack, queen, or king) from a well-shuffled 52-card deck of playing cards is:

    a) 3/52   b) 1/2197   c) 1/13   d) 3/13   e) 3/2197

12. The probability of selecting a heart or a 10 from a well-shuffled 52-card deck of playing cards is:

    a) 4/13   b) 1/51   c) 17/52   d) 15/52   e) none of the above

13. If we are dealing with two mutually exclusive and exhaustive categories and $p(A) = 0.25$, then $p(B)$ equals:

    a) .25   b) .50   c) .75   d) 1.00
    e) insufficient information to answer question

14. The formulation $p(A|B) = p(A)$ shows that the events are:

    a) exhaustive   b) mutually exclusive   c) occurring jointly
    d) independent   e) none of the above

15. We toss a pair of dice. The probability of obtaining a 5 on the first die and a 6 on the second die is:

    a) 1/6   b) 1/3   c) 1/35   d) 1/18   e) 1/36

16. We toss a pair of dice. The probability of obtaining a sum equal to 11 is:

    a) 1/6   b) 1/3   c) 1/35   d) 1/18   e) 1/36

17. Sampling with replacement: Two cards are drawn at random from a 52-card deck of playing cards. The probability of selecting a heart and a king is:

    a) 52/2704   b) 39/2704   c) 16/52   d) 52/2652   e) 39/2652

18. Mutually exclusive events are:

    a) never independent   b) always related   c) dependent
    d) all of the above   e) none of the above

19. When events are dichotomous and mutually exclusive, $p(A|B)$ equals:

    a) $p(B|A)$ b) $1.00 - [p(A) + p(B)]$ c) zero
    d) all of the above   e) none of the above

20. In a single toss of a pair of dice, the probability that the sum will be an odd number and that a four will appear on at least one die is:

    a) 1/6   b) 5/36   c) 18/36   d) 1/9   e) none of the above

21. For nonindependent events, $p(B)p(A|B)$ equals:

    a) $p(A)$ b) $p(B)$ c) $1.00 - p(A)$ d) $1.00 - p(B)$ e) $p(A)p(B|A)$

22. When $p(B|A) \neq p(B)$, the events are:

    a) not independent   b) independent   c) exhaustive   d) nonrandom
    e) none of the above

23. When one is sampling without replacement, $p(A$ and $B)$ equals:

    a) $p(A)p(B)$ b) $p(B|A)$ c) $p(A|B)$ d) $p(A) + p(B)$ e) $p(A)p(B|A)$

24. When one is sampling without replacement, the probability of drawing a queen on each of two draws from a 52-card deck of playing cards is:

   a) 16/2652   c) 12/2652   c) 12/2704   d) 7/52   e) 8/52

25. Given $\mu = 50$, $\sigma = 5$, the probability of selecting at random an individual with a score of 40 or less is:

   a) .45   b) .05   c) .48   d) .975   e) none of the above

26. Given $\mu = 20$, $\sigma = 2$, the probability of selecting at random an individual with a score as rare or unusual as 17 is:

   a) .93   b) .07   c) .87   d) .13   e) .43

27. Given $\mu = 50$, $\sigma = 5$, the probability of selecting at random an individual with a score of 53 or less is:

   a) .73   b) .23   c) .27   d) .47   e) none of the above

28. On the basis of chance, the probability of obtaining a case that falls in the range between $z = -.50$ and $z = -1.00$ under the normal curve is:

   a) 5 in a hundred   b) 15 in a hundred   c) 20 in a hundred
   d) 34 in a hundred   e) 53 in a hundred

29. Sixty percent of the students at a given university come from families with annual incomes of $50,000 or more. Ninety percent of the students at this university have group Rorschach scores that are in the range of "well-adjusted" individuals. If family income and personality adjustment are uncorrelated, approximately what percentage of students are "rich neurotics"?

   a) 6   b) 10   c) 30   d) 40   e) 54

Answers: (1) e; (2) a; (3) b; (4) e; (5) c; (6) c; (7) e; (8) c; (9) d; (10) c; (11) d; (12) a; (13) c; (14) d; (15) e; (16) d; (17) a; (18) d; (19) d; (20) a; (21) e; (22) a; (23) e; (24) b; (25) e; (26) d; (27) a; (28) b; (29) a.

11

Introduction to Statistical Inference

CONCEPTUAL OBJECTIVES

1. What is the purpose of sampling? What is the relationship of a sample to the population?

2. Define and give the function of sampling distributions.

3. State the relationship between the binomial distribution and the concept of sampling distribution.

4. In testing hypotheses, what are the two possible reasons for a sampling outcome? What is the relationship between the probability of the obtained results and chance factors?

5. Specify the two cutoff points commonly used as the basis for inferring the operation of nonchance factors. Define the alpha and the beta levels.

6. What are the null and alternative hypotheses and what is their role in testing statistical hypotheses? Distinguish between directional and nondirectional alternative hypotheses.

7. In terms of indirect proof, state the relationships between accepting and rejecting the null and alternative hypotheses.

8. What are the two types of potential errors when rejecting or accepting the null hypothesis?

CHAPTER REVIEW

Those scientists conducting research in any one of the behavioral sciences find themselves involved with populations in which it is impossible or impractical to determine parameters.

Consider the following questions: In view of the evidence linking heart disease to high levels of animal fat in the diet, what percentage of adult Americans changed their dietary habits to include a greater proportion of foods that contain fats derived from vegetable and marine sources? On a worldwide basis, has the accelerating threat of AIDS altered the sexual practices of sexually active individuals? If so, what changes have occurred? Of the growing number of homeless people who seek refuge in the urban centers of the country, what percentage are emotionally disturbed individuals who might best be helped by mental health professionals?

An attempt to survey the entire target populations to obtain answers to these questions would be needlessly time-consuming and enormously expensive, even assuming that such censuses would be within the realm of possibility. Rather than attempting to determine parameters directly, behavioral scientists have adopted alternative methods that involve estimating parameters from samples drawn from the target populations.

Although you are already acquainted with the concept of sampling, in this chapter we shall explore the use of probabilities in generalizing from a sample to a population, the overall idea of how hypotheses concerning a population are tested from sample statistics, and the errors that might be made in deciding whether or not a hypothesis is valid for a given population.

Fundamental to statistical inference is the practice of drawing a sample from the population being examined. If the population parameters were readily available, there would be no reason to sample, or to hypothesize about the parameters. All questions about the population could be answered merely by looking at the exact values. However, since the parameters of most populations cannot be easily measured, we must rely on the techniques of sampling.

As you should expect, there is some inaccuracy when estimating parameters from a sample. If you were to take ten samples from a population, the chances are that you would obtain ten slightly different means and standard deviations. Perhaps one or two of your sample statistics would differ sharply from the others.

Suppose we were able to draw all possible samples in which $N = 15$ from a population. We could then calculate what is called the sampling distribution of some statistic in which we were interested. Formally defined, a sampling distribution is a theoretical probability distribution of the possible values of some sample statistic that would occur if we were to draw all possible samples of a fixed size from a given population.

Since it is just as impractical to draw all possible samples of a given size from a population as it is to attempt to determine the parameters of that population if it is of vast size, the word "theoretical" appears in the definition of a sampling distribution. Just as the population parameters are usually impractical to determine, so also is the sampling distribution of a statistic. However, it is possible to understand the concept of a sampling distribution by selecting an example in which the population size is so small that we can specify the outcomes of all possible samples of a given size selected from that population. Perhaps we could draw 300 samples in which $N = 15$ from a population. The 300 means from the 300 samples would approximate the sampling distribution of the mean for that population. Similarly, the 300 standard deviations would approximate the sampling distribution (given that $N = 15$) of the standard deviation for that population.

If we were to calculate the standard deviation of our 300 sample means, we would be computing what is called the standard error of the mean. The symbol $s_{\bar{x}}$ designates the standard error of the mean. If $s_{\bar{x}} = 24$, we would know the standard deviation of the 300 sample means was 24.

If the N in our samples were 100 instead of 15, we might expect the standard error of the mean to be smaller than 24. You can reason that the larger the N in each of the samples is, the smaller the expected standard error of the mean should be. Following this logic, we might predict an even smaller standard error of the mean if $N = 500$, rather than 100 or 15.

If we are dealing with a situation in which all observations must fall into one or the other of two possible classes, we have an instance of the binomial case. Through the use of the binomial sampling distribution, we can determine the probability of occurrence of a single outcome or of a *combination* of outcomes.

Let us turn to an example with two mutually exclusive and exhaustive possibilities for each of the outcomes. With this stipulation, we can use the binomial sampling distribution. In this example, we select gender as our category with two mutually exclusive and exhaustive possibilities, which are, of course, female and male. The question we pose for ourselves is "From a sample of ten individuals selected at random from the population of undergraduate physics majors, what is the likelihood that we shall obtain only two females?"

To answer this question, we consult Figure 11.2 in the text and see that the probability of selecting exactly two females when $N = 10$ is 0.044. The probability of selecting two or fewer = 0.055. On the basis of this evidence, are we willing to conclude that females are in the minority among physics majors?

What if we had obtained no females? What is the probability that this event could have occurred by chance? Again, consulting Figure 11.2 in the text, we see that the probability is .001.

At what point can we be reasonably confident that our sample does *not* represent chance variation? In other words, at what point do we feel that the probability that is associated with a specific outcome is so small that we reject the possibility that it could reasonably have occurred simply by chance? Most social scientists have adopted one or the other of the following two probability levels as the cutoff point between a chance or nonchance occurrence:

1. .05 significance level. Some investigators will conclude that if the event would occur by chance 5% of the time or less, then the event can be attributed to nonchance factors.

2. .01 significance level. Other investigators will conclude that if the event would occur by chance 1% of the time or less, then the results are caused by nonchance factors.

The significance level, whether it be .05 or .01, set by the experimenter is referred to as the alpha (α) level. If a scientist chooses a .05 significance level for an experiment, $\alpha = .05$.

We should note that the choice of significance level is arbitrary. Scientists who are more conservative about their findings will tend to select the .01 level. This is to lower the risk of a Type I error, i.e., lower the chance of claiming a significant effect of an experimental variable when there may be, in fact, no effect. Other scientists feel that setting such a stringent criterion for rejecting the null hypothesis is like throwing out the baby with the bath water. In other words, accepting a false null hypothesis as true also carries risks. For one, promising lines of research may be abandoned because of a Type II error, namely, failing to reject a false null hypothesis.

There is an important point to keep in mind that we have not yet mentioned in this discussion: When determining p-values to test for the significance level of an event, we must remember that we might be dealing with a two-tailed distribution and that the p-value should be the probability of obtaining an outcome as rare as *or rarer than* the one in question. If we want to find the p-value corresponding to an event as rare as the occurrence of two females and eight males, we must also consider the possible occurrence of one female and nine males and of ten males. Once we find the combined p-values of these possible outcomes, we must then double it, for we are dealing with a two-tailed distribution. In other words, we are questioning the rarity of selecting a sample as extreme as or more extreme than a sample of two females and eight males. As well as the combined probabilities of the three outcomes we have mentioned, we must also include the combined probabilities of the three outcomes at the other tail of the distribution. These outcomes are two males and eight females, one male and nine females, and zero males and ten females. Since the one-tailed p-value is .055, the two-tailed p-value is .110.

Assuming that $\alpha = .05$, we can assert that an outcome as rare as two females and eight males is not sufficiently rare to rule out chance as possible. Therefore, we should not conclude our results to be attributable to nonchance factors, such as gender difference in preferences.

Our next question is, "How do researchers go about proving a hypothesis they might hold?" Our first reminder here is that a hypothesis can never be proven; at best it can be supported. Support consists of the rejection of the null hypothesis. Let's look at an example.

Suppose that an educator suspects that there are more male than there are female students using the school athletic facilities. By taking a sample of the students on the tennis courts at a given time, she hopes to gain support for her contention. At 4:00 Thursday afternoon she proceeds to the tennis courts to find four females and six males on the courts. From her sample, can the educator conclude that more males use the tennis facilities? The one-tailed p-value is .377. Using $\alpha = .05$, we can claim no significance for the occurrence of such an outcome. In other words, the fact that the educator found four females and six males on the tennis courts does not prove her hypothesis that more males use the courts. If the educator is perseverant, she will return for a larger sample or obtain several more small samples. With more data, she may be prepared to state more conclusively whether or not more males use the school facilities.

Let us describe this microcosmic experiment in other terms. A scientist wishing to test the hypothesis that more male students used the athletic facilities would set up two mutually exclusive hypotheses: the null hypothesis (H_0) and the alternative hypothesis (H_1). H_0, or the null hypothesis, must specify a value for one of the population parameters, whereas H_1, or the alternative hypothesis, must claim some value other than the one mentioned in the null hypothesis.

In our example, the statement of H_0 would be $P \geq Q$ or $P \geq 1/2$, in which P refers to the proportion of females and Q to the proportion of males. In other words, the proportion of females using the athletic facilities is equal to or greater than $1/2$. The alternative hypothesis, H_1 in this case, can be phrased in a directional way: $P < Q$, where P represents the probability of the occurrence of females. If we reject H_0, we assert H_1.

The alternative hypothesis, while it must offer a value for the population parameter other than the one suggested in the null hypothesis, may be either directional or nondirectional. Since the educator is postulating a higher frequency of males, the alternative hypothesis is directional in our example. However, if the alternative hypothesis had been merely that an unequal number of female and male students used the facilities, the statement of H_1 would have been $P \neq Q \neq 1/2$. This is a nondirectional hypothesis. Two-tailed analyses must be made with nondirectional alternative hypotheses, whereas one-tailed analyses must accompany directional alternative hypotheses. So, we would calculate a one-tailed p-value with a directional H_1.

Since our data do not allow us to reject H_0, we fail to reject the null hypothesis. That is, we have not demonstrated a statistically significant difference between the number of males and the number of females using the school athletic facilities. We did not disprove the null hypothesis.

As you can see, our chances of error are greater than we might like to admit. Either we can reject the null hypothesis when it is actually true (Type I error) or we can fail to reject the null hypothesis when it is actually false (Type II error). The educator might replicate her experiment with the possibility in mind that she had made a Type II error with her first sample of data.

SELECTED EXERCISES

1. Why are samples taken from populations?

2. Can there be more than one sample from a single population?

3. What is a sampling distribution? Why is the concept of a sampling distribution important?

4. What is the purpose of models? Give two examples of models.

5. What is the relationship between the level of significance and a Type I error? and a Type II error?

6. What is the probability that a student will correctly guess the answers to 70% or more of the items on a True-False examination, if there are ten items?

7. A manufacturer knows that 50% of his products will be defective. In a random sample of 10, what is the probability that:

a) all will be defective?
b) exactly one will be defective?
c) less than two will be defective?

Answers:

1. In order to obtain estimates of parameters or to test research hypotheses whenever it is impossible or impractical to test an entire population.

2. Yes. It is not unusual to take many samples from a single population, often at different time intervals. Example: public opinion polls.

3. It is a hypothetical distribution of a sample statistic that would be obtained if all possible samples of a given size were selected from a population. Sampling distributions are at the very core of inferential statistics and decision making in the sciences, social sciences, economics, and business.

4. In statistics, a mathematical model is usually an idealized distribution of a variable or a statistic that provides expected results as a reference against which the obtained results may be evaluated. The normal curve and binomial distribution are two examples of such models.

5. The more demanding the α-level (i.e., the lower the probability value defining α) is, the lesser the probability of making a Type I error and the greater is the likelihood of making a Type II error—failing to reject a false H_0. The less demanding the α-level (i.e., the higher the probability value defining α) is, the greater is the likelihood of making a Type I error—and the smaller the likelihood of a Type II error.

6. In the two-category case in which $p = q = {}^1/_2$, the binomial coefficients (Table 1 in the appendix to the text) may be used to find the solution. To find the probability of 70% or more (i.e., 7 or more when $N = 10$), sum the coefficients of 7 through 10 and divide by 1024. Thus, $p = (120 + 45 + 10 + 1)/1024 = 0.172$.

7. This is again a two-category variable in which $p = q = {}^1/_2$.
   a) The coefficient for 10 out of 10 is 1. When $N = 10$, the sum of the coefficients is 1024. Thus, $p = 1/1024 = .001$.
   b) The coefficient for exactly 1 is 10. Thus, $p = 10/1024 = .01$.
   c) This includes the events 1 and 0. These coefficients are 10 and 1, respectively. Thus, $p = (10 + 1)/1024 = .011$.

SELF-QUIZ: TRUE-FALSE

Circle T or F.

T   F   1. Most statistical investigations study a sample rather than the entire population.

T   F   2. A binomial population has many categories.

T   F   3. If we reject a true null hypothesis, we make a Type I error.

T   F   4. If $\alpha = .05$, there is a 5% chance that we will make a Type I error when we reject H_0.

T   F   5. If $\alpha = .05$, $\beta = .95$.

T   F   6. It is accurate to describe a binomial distribution as a model with known mathematical properties that is used to describe certain sampling distributions.

T   F   7. When we draw a large number of samples from a known population, we often find that many sample means differ from the population mean.

T   F   8. With $\alpha = .05$ we are more willing to risk a Type I error than with $\alpha = .01$.

T   F   9. The alternative hypothesis always states a specific value.

T   F   10. If the null hypothesis is true, the probability of making a Type II error equals zero.

T   F   11. Deviations above and below the parameter count equally in a one-tailed test.

T   F   12. Even though we frequently use samples, it is always *possible* to study all the members of a given population.

T   F   13. Since populations can rarely be studied, we are interested in making inferences about samples.

T   F   14. We use models such as the normal curve to describe sampling distributions.

T   F   15. It is possible to estimate the true proportion of heads and tails characteristic of a particular coin from a sample of that coin's "behavior."

T   F   16. The sampling distribution of probabilities for a population consisting of two mutually exclusive and exhaustive categories is given by the normal distribution.

T   F   17. When we toss a coin ten times, we expect to obtain an equal number of heads and tails approximately 50% of the time.

T   F   18. Employing $\alpha = .01$, we reject H_0. If we employed $\alpha = .05$, we would also reject H_0.

T   F   19. Employing $\alpha = .05$, we obtain $p = .02$. We would conclude that the results were due to nonchance factors.

T   F   20. The null hypothesis can never be proved.

T   F   21. If H_1 is directional, we must calculate a two-tailed p-value.

T   F   22. Suppose a researcher finds significance at the .01 significance level. He or she then concludes that nonchance factors are operating.

T   F   23. In most two-category populations, the true values of P and Q cannot be known.

Answers: (1) T; (2) F; (3) T; (4) T; (5) F; (6) T; (7) T; (8) T; (9) F; (10) T; (11) F; (12) F; (13) F; (14) T; (15) T; (16) F; (17) F; (18) T; (19) T; (20) T; (21) F; (22) T; (23) T.

SELF-TEST: MULTIPLE-CHOICE

1. The population of all possible outcomes resulting from tossing a pair of dice is:

    a) finite   b) very large   c) relatively difficult to know conclusively
    d) unlimited   e) none of the above

2. Populations:

    a) can rarely be studied exhaustively   b) may be estimated from samples
    c) are often hypothetical   d) may be unlimited   e) all of the above

3. If the voting preferences of 100 registered voters who voted in the previous election are studied, our primary interest is in:

a) determining how they will vote
b) determining voting preferences of all registered voters
c) estimating voting preferences of individuals who will vote
d) all of the above
e) none of the above

4. If we drew a large number of samples from a known population, we would not be surprised to discover:

a) some differences among the values of the sample statistics
b) a distribution of sample statistics around some central value
c) that many sample means differ from the population mean
d) all of the above
e) none of the above

5. The appropriate mathematical model for describing the sampling distribution of outcomes in a coin-tossing experiment is:

a) the normal curve
b) the binomial distribution in which $P = Q$
c) the binomial distribution in which $P \neq Q$
d) the null hypothesis
e) the alpha (α) level

Given the following theoretical frequency distribution of outcomes with a two-category variable, $N = 5$, $P = Q = \frac{1}{2}$, answer Multiple-Choice Problems 6 through 10.

Number of Ways of Obtaining Outcomes in the A Category

All in A	4 of 5 in A	3 of 5 in A	2 of 5 in A	1 of 5 in A	0 of 5 in A
1	5	10	10	5	1

6. The probability of obtaining exactly four events in the A category is approximately:

a) .31   b) .03   c) .016   d) .16   e) .19

7. The probability of obtaining two or more events in the A category is approximately:

a) .50   b) .81   c) .03   d) .47   e) .63

8. The probability of obtaining a result as rare as one event in the A category is approximately:

a) .19   b) .06   c) .38   d) .31   e) .03

9. The probability of obtaining a result as rare as 3 out of 5 in the A category is approximately:

a) .31   b) .50   c) .72   d) .63   e) 1.00

10. The probability of obtaining all events in the A category is approximately:

a) 1.00   b) .63   c) .17   d) .03   e) none of the above

11. The statement, "The obtained result would have occurred by chance 5% of the time or less," employs:

a) $\alpha = .05$ b) the 5.00% significance level   c) the .05 significance level
d) all of the above   e) none of the above

12. The difference between setting $\alpha = .05$ and $\alpha = .01$ is:

   a) with $\alpha = .05$ we are more willing to risk a Type I error
   b) with $\alpha = .05$ we are more willing to risk a Type II error
   c) $\alpha = .05$ is a more "conservative" test of H_0
   d) with $\alpha = .05$ we are less willing to risk a Type I error
   e) none of the above

13. In a carefully conducted coin-tossing experiment testing $H_0: P = Q = \frac{1}{2}$, we obtain a p-value of .02. Using $\alpha = .05$, we would conclude:

   a) the coin is definitely biased
   b) the coin is probably not biased
   c) the coin is definitely not biased
   d) the coin is probably biased
   e) insufficient information to draw any conclusions

14. In a coin-tossing experiment testing $H_0: P = Q = \frac{1}{2}$, we obtain a p-value of .50. Using $\alpha = .05$, we would conclude:

   a) we cannot reject H_0 As probability is much higher than α
   b) we have disproved H_0
   c) we have proved there is no bias
   d) H_0 is probably false
   e) the coin is extremely well balanced

15. Which of the following cannot be H_0?

   a) the population means are equal
   b) $P = Q = \frac{1}{2}$
   c) $P = \frac{1}{4}, Q = \frac{3}{4}$
   d) the sample means are equal
   e) the difference in the population means from which the sample was drawn is 3.85 points

16. The rejection of H_0 is always:

   a) direct   b) based on the rejection of H_1 c) indirect
   d) based on the direct proof of H_1 e) none of the above

17. In a ten-item true-false examination, the probability that an unprepared student will obtain all correct by chance is approximately:

   a) 1.000   b) .001   c) .202   d) .500   e) .037

Answers: (1) d; (2) e; (3) c; (4) d; (5) b; (6) d; (7) b; (8) c; (9) b; (10) d; (11) d; (12) a; (13) d; (14) a; (15) d; (16) c; (17) b.

12

Statistical Inference and Continuous Variables

BEHAVIORAL OBJECTIVES

Conceptual Objectives

1. State the similarities between the distribution of a sample and the distribution of the sample statistic. What is the relationship between the size of N, the shape of the population, and the properties of the distribution of sample means?

2. Given that the parameters of a population are known, describe the procedure for determining the probability of obtaining a specific sample mean. Define critical regions and their relationship to statistical hypotheses.

3. What are point estimates? What is their function for populations with unknown parameters? Distinguish between biased and unbiased estimates of the parameters.

4. What distributions permit the testing of hypotheses with samples drawn from normally distributed populations in which σ is unknown? Given the z-statistic, write the ratio for the t-statistic. Define and calculate degrees of freedom.

5. Specify the various characteristics of the t-distributions. What are critical values?

6. Distinguish between point estimations and interval estimations.

7. State the relationships among critical regions, confidence limits, confidence intervals, and the decision concerning the hypothetical population parameter.

8. What is the rationale for using a test of significance for correlation coefficients?

Procedural Objectives

1. Using Table A in the appendix to the text and the formula for z-scores, calculate the probability of a mean when the population parameters are known.

2. Given the values of $\bar{X}$, μ_0 and s, compute the value of the t-ratio. In formally setting up the problem, state the null hypothesis, the alternative hypothesis, the significance level, the degrees of freedom, the critical region, and the decision concerning H_0. Familiarize yourself with Table C in the appendix to the text.

3. Given the appropriate information, calculate confidence limits.

4. Using Table F in the appendix to the text, conduct a test of significance for Pearson r.

5. Using Table G in the appendix to the text, test for the significance of r_s.

CHAPTER REVIEW

To begin Chapter 12, we shall continue with the subject of sampling distributions. As you should know from the preceding chapter, if the focus of our attention were the mean and we wished to consider only sample sizes of 4, we would draw all possible samples having $N = 4$ from the population. Once we had calculated the mean for each sample, the resulting distribution of mean scores would be the sampling distribution of means. If you would like to broaden your knowledge of this area, there are three characteristics of sampling distribution of which you should be aware.

1.   The mean of the sampling distribution will not vary with a change in sample size. For instance, if your overall mean from the sampling distribution of means was 10 when $N = 4$, it should remain 10 whether you increase your sample sizes to 7 or decrease them to 3. The mean of a sampling distribution of means is equal to the population mean.

2.   With an enlargement of sample size in a sampling distribution of means, the dispersion of sample means will become less. The larger the N is, the more compact the distribution of sample means is. As N increases, $s_{\bar{X}}$ decreases. If $s_{\bar{X}}$ equals 8 with the sample size of 10, $s_{\bar{X}}$ will be less than 8 if we increase N to 20.

3.   When taken from a normally distributed population, the distribution of sample means will also be bell-shaped. This tendency toward a bell-shaped curve occurs even when the scores in the general population are skewed.

From these three points comes the central limit theorem, which states: If random samples of a fixed N are drawn from any population, as N becomes larger, the distribution of sample means approaches normality with the overall mean approaching μ, the variance of the sample means being equal to σ^2/N, and the standard error equaling $\sigma/\sqrt{N}$. The central limit theorem can also be stated symbolically as follows:

$$\sigma_{\bar{X}}^2 = \frac{\sigma^2}{N}$$

and

$$\sigma_{\bar{X}} = \frac{\sigma}{\sqrt{N}}$$

Let us reflect on a situation that actually does not happen too frequently, a situation in which we know the population parameters, μ and σ. We shall examine the procedure for testing a statistical hypothesis when we know the values for μ and σ.

Suppose a dog trainer would like to compare the sample mean of scores for the dogs she is currently training to the population mean score of all the dogs to which she has ever given the obedience training test. The question she wants to answer is, "What is the probability that the sample mean will exceed a score of 65?"

Before we can begin our calculations, we must know the sample statistics, the population parameters, and the z-score formula we must use. We can then transform the sample mean into a z-score to determine areas under the normal or bell-shaped curve. The formula we must use to find the z-score for our sample mean is:

$$z = \frac{\bar{X} - \mu_0}{\sigma_{\bar{X}}}$$

where μ_0 = the value of the population under H_0, and

$$\sigma_{\bar{X}} = \frac{\sigma}{\sqrt{N}}$$

What is the statement of the null hypothesis? Given that $N = 25$, $\mu = 72$, and $\sigma = 12$, $H_0{:}\mu = \mu_0 = 72$. Substituting the appropriate quantities into the z-score formula, we find

$$z = \frac{65 - 72}{2.4}$$

The value of z corresponding to 65 is -2.92. From column C of Table A in the appendix to the text, we see that only .0018 of all the means would fall below a z of -2.92.

Now we must determine whether a sample having a mean of 65 could reasonably have been drawn from the population with a mean of 72. To do this, we must formally state the problem in statistical terms:

$H_0{:}\mu = \mu_0 = 72$

$H_1{:}\mu \neq \mu_0 \neq 72$

Significance level: $\alpha = .01$

Critical region for rejection of H_0: $|z| \geq |z_{.01}|$

You should be acquainted with all of the preceding terms with the exception of the critical region. The critical region is that portion of area under the curve that includes those values of a statistic that lead to the rejection of H_0. Exactly how much of the two tails of a distribution is contained in the critical region is established by the value of α. An α level of .05 would include greater area than an α of .01. If a z-score lies in the critical region, we must reject the null hypothesis.

In order to decide whether or not to reject or accept the null hypothesis, we must determine if a z-score of -2.92 lies in the critical region. A z-score greater than 2.58 or more than -2.58 (i.e., a z with an absolute value of 2.58 or greater) would have an area associated with it that would place it in the critical region. From this information, we should realize that our z-score of -2.92 does lie in the critical region. We must reject H_0. In other words, a sample with a mean of 65 is not likely to have been drawn from a population with $\mu = 72$ and $\sigma = 12$. In effect, we are affirming

$$H_1{:}\mu \neq \mu_0 \neq 72$$

As we have repeated numerous times, rarely do we know the population parameters as we did in the preceding example. However, with some alterations, we can still use the same logic in cases in which the parameters are unknown as we did in the problem we just completed. To do this, we must estimate our population parameters, μ and σ. When we estimate these values from samples, they are known as point estimates. The formula most frequently used to estimate the standard error of the mean is:

$$\text{Est } \sigma_{\bar{X}} = s_{\bar{X}} = \sqrt{\frac{SS}{N(N-1)}}$$

The sample mean provides the point estimate of μ.

Let us now explore the procedure for obtaining a z-score type of statistic. This statistic, analogous to z-score, will enable us to decide whether to accept or reject the null hypothesis. When parameters are not known, the normal curve is not the appropriate sampling distribution. The appropriate distribution is more spread out than the normal distribution. However, thanks to a statistician named Gossett, there is a statistic known as the t-statistic that can be used in the same manner as the z-score. The t-statistic is used with the t-distributions. Just as the z-score is applied to the normal distribution, so is the t-statistic applied to t-distributions.

The formula for the t-ratio is

$$t = \frac{\bar{X} - \mu_0}{s_{\bar{X}}}$$

You may have noted that instead of referring to a single t-distribution, we mentioned the plural, t-distributions. Actually, there is a family of t-distributions rather than a single distribution. Which t-distribution you should use in a particular problem depends on the number of degrees you have.

The term "degrees of freedom" refers to the number of scores that are free to vary once you have placed restrictions on the data set. The more restrictions that are placed on a data set, the fewer the degrees of freedom there are, because the scores are not so free to vary. If we were to restrict a sample of five scores by saying that the mean must equal 19, we have placed one restriction on our data. With a single restriction, the number of degrees of freedom equals $N - 1$. If there were two restrictions, the degrees of freedom would equal $N - 2$. The t-distribution for a given problem depends on the number of degrees of freedom. The larger the number of degrees of freedom is, the more like the normal distribution the t-distribution becomes.

Just as with z-scores, we must consult a table (Table C in the appendix to the text) to make use of the t-ratio. However, instead of areas under the curve, critical values are listed. To use Table C, you must know the number of degrees of freedom and the significance level of your problem. The values you consult are critical values, those values bounding the critical regions that are associated with various significance levels. For instance, if you have 26 degrees of freedom and a significance level of .05 for a two-tailed test, the critical value in which you are interested is |2.056|. This means that if the t-ratio you obtain is greater than or equal to |2.056|, you may reject H_0. In other words, if your t-ratio exceeds the critical value listed in Table C, you must reject H_0.

On the other hand, if your t-ratio is less than the critical value, you may not reject the null hypothesis.

We may also use the t-distribution to obtain interval estimates. An interval estimation is the statement of a range of values within which we think the parameter falls. By adopting an interval estimate, we are stating that we feel the population parameter falls between two values.

The statement of the range of the interval estimate can be made with varying degrees of confidence. Obviously, if the interval estimate includes the entire range of the population distribution, we can be 100% confident the μ is contained within our interval. However, if we establish our confidence interval using an $\alpha = .05$, we have a 95% confidence interval, as opposed to a 100% confidence interval. Similarly, if $\alpha = .01$, we would be dealing with a 99% confidence interval. The two values defining a confidence interval are called confidence limits. It stands to reason that the two values, or confidence limits that bound the 99% confidence interval of a distribution allow a larger range of values than do the confidence limits of the 95% confidence interval of the same distribution. Naturally, the greater the size of the interval is, the more confident we are that the population parameter lies within the confidence limits of the interval.

To determine the limits of the 95% confidence interval, you should use the following two formulas:

$$\text{Upper limit } \mu_0 = \bar{X} + t_{.05}(s_{\bar{X}})$$

and

$$\text{Lower limit } \mu_0 = \bar{X} + t_{.05}(s_{\bar{X}})$$

Note that the quantity represented by the symbol $t_{.05}$ is the critical value at the .05 significance value for a specific number of degrees of freedom. As you know, the critical value changes with the number of degrees of freedom. If we wished to obtain the confidence limits of the 99% confidence interval, we would substitute $t_{.01}$ for $t_{.05}$ in the two formulas for the confidence limits.

Since we are using z- and t-scores to test for significance in rejecting or accepting H_0, let us examine one more area in which this same principle is applied, the area of correlation coefficients. When ρ, the population correlation coefficient, is the parameter that we are attempting to estimate from the sample statistics, H_0 is usually stated as $\rho = 0.00$. Of course, the two-tailed H_1 is then $\rho \neq 0.00$.

When H_0 is that α equals 0.00, the appropriate test statistic is t, which is defined as

$$\frac{r \sqrt{N-2}}{\sqrt{1-r^2}}$$

To test the null hypothesis that the population correlation is some value *other than* 0.00, a special form of z-statistic is required. The formula for z is

$$z = \frac{z_r - Z_r}{\sqrt{\dfrac{1}{N-3}}}$$

where z_r = the transformed value of the sample r, and Z_r = the transformed value of the population correlation coefficient specified under H_0. For sample values of r, the corresponding z_r is found in Table F in the appendix to the text. For a given value of r, you should consult Table F to find Z_r. Knowing the z-score of r, you can use Table A to determine the significance for various values of α. If the z-score of r lies in the critical region, you must reject H_0. If the z-score does not fall in the critical region, then you cannot reject the hypothesis that equals the value specified under H_0.

SELECTED EXERCISES

1. Under what conditions will the sampling distribution of the mean be normal?

2. What are some of the differences between the distribution of a sample and the distribution of a sample statistic?

3. If samples are randomly drawn from the *same* population, would we expect the obtained sample means to differ from sample to sample?

4. What happens to the variability of the sample means as we increase the sample size?

5. What does the standard error of the mean indicate?

6. The mean of a population, μ, is 75, and the standard deviation, σ, is 20. Suppose that we draw all possible random samples of 25 cases from this population and calculate the mean of each sample. What will be the standard deviation of this distribution? Is this value considered an estimate? Explain.

7. The mean of a population of scores, μ, is 20, and the standard deviation, σ, is 6:

   a) What is the probability that a single score, selected at random, from this population will be
   i)   above 21?
   ii) above 23?
   iii) between 17 and 23?
   iv) between 19 and 21?

   b) What is the probability that the mean of a random sample of 4 cases drawn from this population will be
   i)   above 21?
   ii) above 23?
   iii) between 17 and 23?
   iv) between 19 and 21?

   c) What is the probability that the mean of a random sample of 36 cases drawn from this population will be
   i)   above 21?
   ii) above 23?
   iii) between 17 and 23?
   iv) between 19 and 21?

8. What is the relationship between level of significance and confidence interval?

9. Why is it usually inappropriate to use z as the test statistic?

10. Is

$$s^2 = \frac{\Sigma(X - \bar{X})^2}{N}$$

   an unbiased estimate of σ^2? Explain. What is the unbiased estimate of σ^2?

11. Is the sample mean an unbiased estimate of the population mean?

12. Is there more than one t-distribution? On what do they depend?

13. Explain the concept *degrees of freedom.*

14. What happens to the form of the t-distribution as the degrees of freedom become infinitely large?

15. What is the critical value of t corresponding to $\alpha = .05$, two-tailed test, when:
    a) $N = 2$ b) $N = 4$ c) $N = 8$ d) $N = 16$ e) $N = 30$

16. What are point estimates? Give two examples.

17. What is interval estimation? How does this differ from point estimation?

18. Explain, in your own words, what is meant by the 95% confidence interval.

19. What distributions do we use to calculate confidence limits if the population variance is not known?

20. A dress manufacturer wanted to find out if a recently purchased machine increased the production output of his employees. The previous mean production was 36 pieces an hour. The mean output of 10 randomly selected employees using the new machine is 42.2 with $s = 6$. Employing $\alpha = .01$, one-tailed test, what did he conclude?

21. A group of 26 women follow a special diet for a one-month period. Their weight losses are recorded, yielding $\bar{X} = 9.5$ pounds and $s = 2.5$ pounds. Find the 95% confidence limits for μ.

22. The chairperson of the speech department at a large university believes that his department is younger than the average. The mean age of the entire faculty is 42.4. Listed here are the ages of 12 members of the speech department. Employing $\alpha = .05$, what do you conclude? (Set this problem up in formal statistical terms, H_0, H_1, etc.)

28	39	36	42
29	32	62	44
30	43	44	49

23. Employing the data in the preceding exercise, find the

   a) 95% confidence limits for μ

   b) 99% confidence limits for μ

24. The vice-president of a certain bank wanted to find out if her depositors had larger bank accounts than average. The mean account for all banks was $375.46. A random sample of 50 of her depositors showed a mean account of $437.50 with a standard deviation of $75.00. What did she conclude?

25. The caddies at the public golf course contend that people tend to tip less at a public course than they do at a private golf course. The mean tip at a private course is $1.00. Listed here are the tips received by ten randomly selected caddies at a public golf course. What is your conclusion?

$0.50	1.50
1.00	0.75
0.75	0.50
1.25	1.00
1.00	1.00

26. Employing the data in the preceding problem, find the

   a) 95% confidence limits for μ

   b) 99% confidence limits for μ

27. The owner of a hydraulic valve company knows that of every 1000 valves his competitor produces, 20 will be defective as a long-term average. in order to compare the efficiency of his production, he randomly selects 15 samples of 1000 valves each and obtains the following number of defectives. At $\alpha = .05$, what does he conclude?

18	16	19	22	21
23	26	21	16	20
17	20	27	18	22

28 Employing the data in the preceding problem, find the 95% confidence limits. Does the interval that you have determined change your answer to Selected Exercise 27?

29. Given the following 50 observations, test

   a) $H_0: \mu = \mu_0 = 25$

   b) $H_0: \mu = \mu_0 = 19$

18	37	15	31	19
33	32	30	34	26
28	15	39	20	10
29	39	5	24	36
22	30	13	23	26
23	18	20	26	25
14	5	41	12	27
19	25	29	15	51
11	38	21	27	20
31	17	25	33	35

30. Employing the data in Selected Exercise 29, find the

    a) 95% confidence limits for μ

    b) 99% confidence limits for μ

31. What is the t-ratio of $r \geq .35$ if the population coefficient is hypothesized to be 0 and $N = 28$?

32. The coefficient of correlation obtained between 19 pairs of scores is .49. Employing $\alpha = .05$, test the hypothesis that the population coefficient is 0.

33. We obtain $r = .68$ between 50 pairs of scores. Employing $\alpha = .01$, test the hypothesis that the population coefficient is .80 or greater.

Answers:

1. The sampling distribution of the mean will approach normality when the underlying distribution of scores is normal. When the underlying distribution is not normal, the distribution of sample means more closely approaches normality as the N in the samples increases.

2. Some of the differences are: the distribution of sample means becomes increasingly leptokurtic as N increases; the standard deviation of the sample means (i.e., the standard error of the mean) becomes smaller as N increases (in other words, the dispersion of sample means decreases); nonnormal distributions take on the characteristics of the normal curve.

3. Yes. This is due to the operation of uncontrolled chance factors.

4. The variability decreases.

5. It is to the distribution of sample means what the standard deviation is to the distribution scores or values of the variable.

6. $\sigma_{\bar{X}} = 20/\sqrt{25} = 4$. Since all possible outcomes have been enumerated, the sampling distribution represents the "true" distribution of sample means. It is not an estimate.

7. a) i) $z = .17, p = .4325$ b) i) $z = .33, p = .3707$
     ii) $z = .50, p = .3085$ ii) $z = 1.00, p = .1587$
     iii) $p = .3830$ iii) $p = .6826$
     iv) $p = .1350$ iv) $p = .2586$

   c) i) $z = 1.00, p = .1587$
      ii) $z = 3.00, p = .0013$
      iii) $p = .9974$
      iv) $p = .6826$

8. The critical values at a given level of significance bracket the confidence interval. Thus, the critical upper and lower values at $\alpha = .05$, two-tailed test, include 95% of the area—the region of acceptance of H_0.

9. It is rare that we know the population variance and standard deviation. For very large samples, z may be substituted for t but little is gained in the way of computational ease.

10. No. To obtain an unbiased estimate of the variance, SS must be divided by degrees of freedom, $N - 1$ in the one-sample case.

11. Yes, as long as the sample is drawn from the population by some random process.

12. Yes. There is a separate t-distribution for each number of degrees of freedom. This is why t-distributions are rarely published but critical values are.

13. Degrees of freedom are the number of values that are free to vary once you have placed restrictions on the data.

14. It approaches the normal curve.

15. a) 12.706   b) 3.182   c) 2.365   d) 2.131   e) 2.045

16. It is the estimation of a parameter based on a single statistic. If we select a sample at random from a population and obtain $\bar{X} = 62$ and $s = 12.3$, these are our point estimates of the mean and standard deviation, respectively.

17. It is the specification of an interval within which the population parameter of interest is presumed to fall.

18. Your answer should *not* state that the parameter has a 95% probability of being within that interval, since the interval is variable, depending on the sample statistics. However, it should contain the thought that repeated representative samples drawn from the population will include the parameter about 95% of the time.

19. t-distributions

20. $t = 3.10$, df $= 9$

21. 8.47 – 10.53

22. $\sum X = 478$ $t = \dfrac{39.83 - 42.40}{2.83}$

    $\sum X^2 = 20{,}096$

    $\sum X^2 = 1055.67$ $t = -.908$, df $= 11$

    $s_{\bar{X}} = \sqrt{\dfrac{1055.67}{132}} = 2.83$

23. a) $39.83 \pm (2.83)(2.201) = 39.83 \pm 6.23 = 33.60$ to 46.06

    b) $39.83 + (2.83)(3.106) = 39.83 + 8.79 = 31.04$ to 48.62

24. $t = \dfrac{437.50 - 375.46}{10.71} = 5.792$, df $= 49$

25. $\sum X = 9.25$ $t = \dfrac{.9250 - 1.0000}{.0990}$

    $\sum X^2 = 9.4375$, SS $= .8812$, $t = -.758$, df $= 9$
    $s_{\bar{X}} = .0990$

26. a) $9.25 \pm (.009)(2.262) = 9.25 \pm .2239 = .7011$ to 1.1489
    b) $9.25 \pm (.099)(3.25) = 9.25 \pm .3218 = .6032$ to 1.2468

27. $\sum X = 306$ $t = \dfrac{20.40 - 20.00}{.85}$

   $\sum X^2 = 6.394$, SS $= 151.6$, $t = .471$, df $= 14$
   $s_{\bar{x}} = .85$

28. a) $20.4 \pm (.85)(2.145) = 20.4 \pm 1.823 = 18.58$ to 22.22
    b) No

29. a) $t = \dfrac{24.84 - 25.00}{1.37} = -.117$, df $= 49$

    b) $t = \dfrac{24.84 - 19.00}{1.37} = 4.263$, df $= 49$

30. a) $24.84 \pm (1.37)(2.01) = 24.84 \pm 2.75 = 22.09$ to 27.59
    b) $24.84 \pm (1.37)(2.28) = 24.84 \pm 3.67 = 21.17$ to 28.51

31. $t = 0.35\sqrt{28 - 2}/\sqrt{1 - .1225} = 1.785/.937 = 1.91$

32. $t = 0.49\sqrt{19 - 2}/\sqrt{1 - .2401} = 2.02/.87 = 2.32$

33. $z = (.829 - 1.099)/\sqrt{1/47} = -.027/.146 = -1.85$

SELF-QUIZ: TRUE-FALSE

Circle T or F.

T   F   1. The t-distributions are normal.

T   F   2. $\hat{s}^2$ is an unbiased estimate of s^2.

T   F   3. When we use t as the test statistic, the normal curve is the model for the sampling distribution.

T   F   4. The variability of the sample means decreases as N increases.

T   F   5. When $\alpha = .01$, the critical region includes 99% of the area.

T   F   6. For any given sample on which we have placed a single restriction, the number of degrees of freedom is 1.

T   F   7. The proportion of area beyond a specific value of t is greater than the proportion of area beyond the corresponding value of z.

T   F   8. The probability of drawing extreme values of the sample mean becomes smaller as N increases.

T   F   9. The mean of any single sample is no more likely to be closer to the mean of the population as sample size increases.

T   F   10. When we do not know the exact values of the parameters μ and σ, we must estimate the test statistic z from sample data.

T   F   11. $s^2 = \dfrac{SS}{N}$ is appropriate only for describing the variability of a sample.

T   F   12. $\hat{s}^2 = \dfrac{SS}{N - 1}$ provides a biased estimate of the population variance.

T   F   13. Since substituting $\hat{s}$ for σ \provides a reasonably good approximation to the sampling distribution of means, we may use the normal curve as the model for our sampling distribution.

T   F   14. With small samples, $\hat{s}$ tends to underestimate σ.

T   F   15. The symbol μ_0 represents a population mean of zero.

T   F   16. If we obtain a negative t-ratio, this means that the true value of the population mean is less than the value hypothesized.

T   F   17. The tabled values for t represent the minimum value of t required for significance at varying α levels.

T   F   18. When we employ samples to estimate parameters, there is no way to determine the amount of error we are likely to make.

T   F   19. Given: the 95% confidence limits for μ are 101 and 107. Employing $\alpha =$ .05, we would reject $H_0{:}\mu_0 = 108$.

T   F   20. Given: the 99% confidence limits for μ are 180 and 185. If we employ $\alpha =$ .05, there is a possibility that we may accept $H_0{:}\mu_0 = 186$.

T   F   21. Given: the 95% confidence limits for μ are 201 and 209. The lower limit for the 99% confidence interval will be a value equal to or less than 201.

T   F   22. The mean of the sampling distribution does not vary with sample size.

T   F   23. A critical region includes all the area in which a sample statistic must lie for rejection of H_0.

T   F   24. For a sample on which we have placed a single restriction, $N - 1$ defines the number of degrees of freedom.

T   F   25. More often than not, $\sigma_{\bar{X}} = \sigma$.

T   F   26. Since sample r's cannot be accurately transformed into z-scores, the normal distribution is of no value in testing hypotheses concerning ρ.

Answers: (1) F; (2) F; (3) F; (4) T; (5) F; (6) F; (7) T; (8) T; (9) F; (10) F; (11) T; (12) F; (13) F; (14) F; (15) F; (16) F; (17) T; (18) F; (19) T; (20) F; (21) T; (22) T; (23) T; (24) F; (25) F; (26) F.

SELF-TEST: MULTIPLE-CHOICE

1. In a normally distributed population with $\mu = 20$, $\sigma = 5$, which of the following sample sizes will yield the smallest variation among sample means? N equals:

   a) 1   b) 5   c) 20   d) 50   e) 250

2. In a normally distributed population with $\mu = 15$, $\sigma = 4$, we draw, with replacement, samples of $N = 2$. Which pair of scores is *least* likely to be selected?

   a) 12, 18   b) 15, 15   c) 16, 14   d) 11, 19   e) 13, 17

3. In a normally distributed population with $\mu = 50$, $\sigma = 10$, we draw samples of $N = 9$. The standard error of the mean is:

   a) .90   b) 1.11   c) 5.56   d) 3.57   e) 3.33

4. In a normally distributed population with an unknown mean, which of the following sample sizes is likely to most closely approximately μ? N equals:

   a) 1   b) 5   c) 20   d) 50   e) 250

5. When μ and σ are known, the form of the sampling distribution of sample means when N is large is normal and:

   a) $\mu_{\bar{x}} = 0$ b) $\mu_{\bar{x}} = \dfrac{\sigma}{N}$ c) $\mu_{\bar{x}} = \dfrac{s}{\sqrt{N-1}}$ d) $\mu_{\bar{x}} = \mu$

   e) cannot answer without knowing the exact value of N

6. In Multiple-Choice Problem 5, $\sigma_{\bar{x}}$ equals:

   a) $\dfrac{s}{N-1}$ b) $\dfrac{\sigma}{N}$ c) $\dfrac{\sigma}{\sqrt{N}}$ d) $\dfrac{s}{\sqrt{N-1}}$

   e) cannot answer without knowing the exact value of N

7. The appropriate statistic for testing hypotheses concerning the value of μ when σ is known is:

   a) $\dfrac{\bar{X} - \mu_0}{\sigma_{\bar{x}}}$ b) $\dfrac{\bar{X} - \mu_0}{s_{\bar{x}}}$ c) $\dfrac{\sigma}{\sqrt{N}}$ d) $\dfrac{s}{\sqrt{N-1}}$

8. The appropriate statistic for testing hypotheses concerning the value of μ when σ is unknown is:

   a) $\dfrac{\bar{X} - \mu_0}{\sigma_{\bar{x}}}$ b) $\dfrac{\bar{X} - \mu_0}{s_{\bar{x}}}$ c) $\dfrac{\sigma}{\sqrt{N}}$ d) $\dfrac{s}{\sqrt{N-1}}$

9. That portion of the area under the curve that includes those values of a statistic that lead to rejection of the null hypothesis is known as

   a) the law of large numbers   b) Student's t-ratio   c) interval estimation
   d) the central limit   e) the critical region

10. Given: $\sigma = 20$, $\bar{X} = 210$, $N = 16$, the appropriate statistic for testing $H_0{:}\mu_0 = 200$ yields a value of:

    a) 1.93   b) 8.20   c) 2.00   d) 2.24   e) 5.00

11. Given: $\mu_0 = 40$, $\sigma = 10$, $\bar{X} = 50$, $z_{.01} = 2.58$. Testing H_0, we would conclude:

    a) reject H_0
    b) fail to reject H_0
    c) the sample mean was not drawn from a population in which $\mu = 40$
    d) $\mu_0 = \mu$
    e) insufficient information to answer question

12. Given: $\bar{X} = 40$, $s = 10$, $\mu_0 = 30$, $N = 10$, the appropriate test statistic is:

    a) t b) z c) $\alpha = .05$ d) $\sigma/\sqrt{N}$
    e) insufficient information to answer question

13. Given: $\alpha = .05$, two-tailed test, $t_{.05} = \pm 2.16$, we obtain a t-ratio of 1.94. Our decision:

    a) reject H_0 b) conclude that H_0 is true   c) fail to reject H_0
    d) none of the above   e) insufficient information to answer question

14. Given: $\alpha = .01$, $t_{.01} = \pm 2.48$, two-tailed test, $\bar{X} = 20$, $s_{\bar{x}} = 3$, and $\mu_0 = 15$. Our decision:

    a) reject H_0
    b) fail to reject H_0
    c) the sample mean is significantly less than the population mean
    d) reject H_0 at $\alpha = .05$ but not at $\alpha = .1$
    e) insufficient information to answer question

15. The estimated standard error of the mean may be obtained from:

   a) $s/\sqrt{N-1}$ b) $\hat{s}/\sqrt{N}$ c) $\sqrt{s^2/(N-1)}$
   d) all of the above   e) none of the above

16. If we were to employ the z-statistic to test H_0 when N is small and σ is unknown, we would:

   a) increase the risk of a Type I error
   b) decrease the risk of a Type I error
   c) increase the risk of a Type II error
   d) both (b) and (c)
   e) none of the above

17. Given: $\alpha = .05$, $\bar{X} = 16$, $s = 4$, $N = 30$, the number of degrees of freedom is:

   a) 17   b) 3   c) 15   d) 31   e) 29

18. Given: $\bar{X} = 20$, $s = 8$, $N = 5$, $\mu_0 = 15$, then $s_{\bar{X}}$ is equal to:

   a) 1.6   b) 4.0   c) 3.6   d) 2.0   e) 2.5

19. When we specify the interval that probably contains the parameter, we are stating

   a) the confidence interval   b) the critical region   c) the α-level
   d) the confidence limits   e) point estimation

20. Given: $z_{.005} = \pm 1.96$, $\bar{X} = 30$, $\sigma_{\bar{X}} = 5$, the lower limit of the interval containing μ is:

   a) 20.20   b) 28.04   c) 15.31   d) 25.00   e) 24.00

21. Given: $t_{.01} = \pm 3.00$, $\bar{X} = 50$, $s_{\bar{X}} = 7$, the upper limit of the interval containing μ is:

   a) 53.33   b) 52.10   c) 57.00   d) 53.00   e) 71.00

22. Employing $z_{.01} = \pm 2.58$, $\bar{X} = 35$, we establish the confidence limits as 30 and 40. We may conclude that:

   a) the probability that $\mu = 35$ is .01
   b) the probability that $\mu = 35$ is .99
   c) the probability that the interval contains μ is .01
   d) the probability that the interval contains μ is .99
   e) none of the above

23. Which of the following is the best illustration of a null hypothesis?

   a) the mean of the population is 100
   b) the mean of the sample is 100
   c) the mean of the population is not 100
   d) the mean of the sample is not 100
   e) all of the above are acceptable null hypotheses

24. As the sample size increases, if the sample standard deviations remain the same, the standard error of the mean will:

   a) increase   b) decrease   c) remain the same
   d) increase at first then decrease   e) must know the sample size

25. t-distribution is:

   a) the same as the normal curve   b) symmetrical   c) skewed
   d) bimodal   e) none of these

26. Suppose you hypothesize that the mean of a population is 50. You draw a random sample of 50 cases from that population. You find that $\bar{X} = 55$ and $s = 14$. What is the value for t?

    a) .07   b) .30   c) 1.0   d) 2.5   e) 3.5

27. If, in a t-test, $\alpha = .05$:

    a) 5% of the time we shall say that there is a real difference, when there really is not a difference
    b) 5% of the time we shall make a correct inference
    c) 95% of the time we shall make an incorrect inference
    d) 5% of the time we shall say that there is no real difference, but in reality there is a difference
    e) 95% of the time the null hypothesis will be correct

28. The standard deviation of a sample distribution of means is called the:

    a) sample standard
    b) standard difference of the sample
    c) standard error of the mean
    d) sample deviation of errors
    e) standard mean of the differences

29. The critical value for a two-tailed z-test with $\alpha = .01$ as the level of significance would be:

    a) ±1.96   b) ±1.64   c) ±2.33   d) ±2.58   e) none of these

30. An estimator is said to be unbiased if:

    a) it approaches nearer and nearer the population value as the sample size increases indefinitely
    b) the average of a large number of sample estimates tends to equal the parameter being estimated
    c) the sample mean is a far better estimator of the central value of the population than the median
    d) the population sample is small
    e) none of the above

31. The term "level of confidence" pertains to:

    a) level of significance   b) hypothesis testing   c) interval estimation
    d) only skewed distribution   e) none of the above

32. In a z-test we have decided to use a two-tailed test with $\alpha = .05$ as our level of significance. If, in fact, the null hypothesis is true, what is the probability of making a Type I error?

    a) .05   b) .025   c) .02   d) zero   e) .10

33. Assume a distribution in which $s = 6$. What is the standard error of the mean based on samples in which $N = 10$?

    a) 167   b) 1.5   c) 3.0   d) 2.0
    e) We must know the value of the mean to answer this question.

34. The difference between means of samples drawn at random from the same population represents:

    a) bias due to the method of measurement
    b) the effects of variation in the independent variable
    c) the effects of the previous experience of subjects on which observations are made

   d) the effects of chance factors
   e) the effects of the confidence limits

35. In order to determine the 99% confidence limits for a mean, which of the following statistics must one know?

   a) the standard error of the difference between means
   b) the standard deviation of the sample
   c) *r* between samples
   d) the true mean of the population
   e) the true mean of the sampling distribution

Answers: (1) e; (2) d; (3) e; (4) e; (5) d; (6) c; (7) a; (8) b; (9) e; (10) c; (11) e; (12) a; (13) c; (14) b; (15) d; (16) a; (17) e; (18) b; (19) a; (20) a; (21) e; (22) d; (23) a; (24) b; (25) b; (26) d; (27) a; (28) c; (29) d; (30) b; (31) c; (32) a; (33) d; (34) d; (35) b.

Statistical Inference with Two Independent and Two Correlated Samples

BEHAVIORAL OBJECTIVES

Conceptual Objectives

1. Distinguish between the sampling distribution of sample means and the sampling distribution of the difference between means. Specify the parameters of the sampling distribution of the difference between means in relation to the population parameters.

2. When can the z-statistic be used with the sampling distribution of the difference between means? State the formula for the unbiased estimates of the standard error, $\sigma_{\bar{x}_1-\bar{x}_2}$ when $N_1 \neq N_2$.

3. If the population standard deviations are unknown, what are the statistics and the formula for testing the hypotheses about the difference between means? Know how to calculate the degrees of freedom that are associated with such a test.

4. Know how to estimate the degree of association (ω^2) between the variables.

5. What are the three assumptions underlying the use of the t-distributions? To test for the homogeneity of variance, you must follow what procedure?

6. What are the three factors reflected by a subject's score on the dependent variable? What is the purpose of using correlated samples?

7. Explain the before-after design and the matched group design. What are the advantages and disadvantages of using matched groups in reference to the standard error of the difference between means? How does the magnitude of r influence the standard error?

8. Given the direct difference method of calculating the Student's t-ratio, identify $\bar{D}$ and $s_{\bar{D}}$. What is the formula for t?

Procedural Objectives

1. Given sample data and using the t-ratio, conduct a test of significance for two independent samples. You should be prepared to list the null hypothesis, the alternative hypothesis, the significance level, the degrees of freedom, the critical region, the value of t, and the decision concerning t.

2. Find the degree of association by calculating estimated ω^2.

3. Determine the F-ratio and refer to Table D in the appendix to the text in conducting a test for homogeneity of variance.

4. Given data from two correlated samples, conduct a test of significance using the direct difference method for calculating the t-ratio. Specify the null hypothesis, the alternative hypothesis, significance level, critical region, value of t, and the decision concerning the null hypothesis.

CHAPTER REVIEW

Perhaps investigators in the behavioral sciences are most often concerned with the possibility that two samples were drawn from the same population. Until this point we have tested hypotheses about only one sample; now our study will switch to two-sample hypothesis testing. Consider the following:

Do individuals exhibiting Type A behaviors experience a greater incidence of cardiovascular disorders—hypertension, heart attacks, strokes—than those classified as Type B?

Do individuals suffering from season affective disorder show improvement when exposed to daily light treatment as compared to control subjects who are exposed to daily sessions of dim illumination?

Researchers addressing themselves to these questions must direct investigations of two-sample hypothesis testing. The samples could be a group of women executives' salaries versus the earnings of a number of their male counterparts, the ages at which a sample of Congonese children walk as compared to a similar sample from the United States, and the consumption levels of a number of American households versus a sample of the consumption levels of European families.

Just as with one-sample cases, our two-sample significance testing will require a null hypothesis, a directional or nondirectional alternative hypothesis, a statistical test (such as the t-statistic), the selection of a significance level, a sampling distribution, and a critical region for our hypothesis testing. With a discussion of the few minor differences in these steps as a result of two-sample problems, rather than one, you should find the procedures readily transferable to the two-sample case. However, in essence our approach will be identical.

When we were dealing with one sample, we described the sampling distribution of the mean as being a distribution of all the values of the mean that would occur if we were to draw all the samples of a population. If N were to equal 4 in one sample, it would equal 4 in another sample.

However, when we are hypothesizing with two samples, our definition of the sampling distribution of sample means changes somewhat. With two samples, not only can sample size vary (N_1 does not have to equal N_2), but also we are considering drawing a large number of *pairs* of samples, rather than just a large number of samples. Remember that the N in these pairs do not have to be equal in size. If we obtain the difference between means in each pair of samples, the resulting distribution is the sampling distribution of the difference between means. As you see, with two samples, our sampling distribution is not a distribution of sample means, but rather, a distribution of the differences between pairs of sample means.

The standard deviation of the sampling distribution of the difference between means is called the standard error of the difference between means and is designated by $\sigma_{\bar{X}_1 - \bar{X}_2}$. Obviously, the symbol reflects the fact that we are dealing with the standard deviation of the distribution of the difference between pairs of means. The standard error of the difference between means is given by the following formula:

$$\sigma_{\bar{X}_1 - \bar{X}_2} = \sqrt{\sigma_{\bar{X}_1}^2 + \sigma_{\bar{X}_2}^2}$$

Just as we can obtain a value for the standard error of the difference between means, so too we can determine the mean of this sampling distribution. The mean ($\mu_{\bar{X}_1 - \bar{X}_2}$) equals the difference between the means of the two populations:

$$\mu_{\bar{X}_1 - \bar{X}_2} = \mu_1 - \mu_2$$

Be certain that you are familiar with the notation for the mean, $\mu_{\bar{X}_1 - \bar{X}_2}$, and for the standard error, $\sigma_{\bar{X}_1 - \bar{X}_2}$, of the sampling distribution of the difference between means.

Since the parameters $\sigma_{\bar{X}_1-\bar{X}_2}$ and $\mu_{\bar{X}_1-\bar{X}_2}$ are usually unknown, we must turn to estimations of these parameters and to the t-statistic for hypothesis testing. To use the t-statistic, we must estimate $\sigma_{\bar{X}_1-\bar{X}_2}$. This estimate of $\sigma_{\bar{X}_1-\bar{X}_2}$ is represented by the notation $s_{\bar{X}_1-\bar{X}_2}$ and the formula

$$s_{\bar{X}_1-\bar{X}_2} = \sqrt{\left(\frac{SS_1 + SS_2}{N_1 + N_2 - 2}\right)\left(\frac{1}{N_1} + \frac{1}{N_2}\right)}$$

If N_1 and N_2 are equal for the two samples, the formula for $s_{\bar{X}_1-\bar{X}_2}$ simplifies to

$$s_{\bar{X}_1-\bar{X}_2} = \sqrt{\frac{SS_1 + SS_2}{N(N - 1)}}$$

The t-ratio application in the two-sample case is identical to its application in problems with single samples. In order to test hypotheses at various significance levels, you must determine the degrees of freedom, the t-ratio, and whether you have a two-tailed or a one-tailed alternative hypothesis. When you have two samples, you must subtract a degree of freedom (df) for each sample. In other words, you have one restriction for each sample. In the two-sample situation, df $= N_1 + N_2 - 2$. If $N_1 = 10$ and $N_2 = 24$, df $= 10 + 24 - 2 = 32$. In consulting Table C, you would be interested in the t-distribution corresponding to 32 degrees of freedom.

Recall that the formula for the t-ratio, one-sample case, is

$$t = \frac{\bar{X} - \mu_0}{s_{\bar{X}}}$$

With two samples $\bar{X}$ is replaced by $(\bar{X}_1 - \bar{X}_2)$ and μ_0 by $(\mu_1 - \mu_2)$. Also, $s_{\bar{X}}$ becomes $s_{\bar{X}_1-\bar{X}_2}$, the estimate for the standard error of the difference between means. As you can surmise, each of the three terms in the t-ratio formula is replaced by the term indicating that we are dealing with the sampling distribution of the difference between means rather than just the sampling distribution of means. If you substitute the appropriate terms in the t-ratio formula, you will have the formula to use in the two-sample case:

$$t = \frac{(\bar{X}_1 - \bar{X}_2) - \mu_1 - \mu_2}{\sqrt{\left(\frac{SS_1 + SS_2}{N_1 + N_2 - 2}\right)\left(\frac{1}{N_1} + \frac{1}{N_2}\right)}}$$

Of course, if $N_1 = N_2$, the denominator can be simplified to

$$\sqrt{\frac{SS_2 + SS_2}{N(N - 1)}}$$

Let's look at an example we first examined in Chapter 5 of this Study Guide (Hsu et al., 1981). Recall that the investigators compared the percentage of body cells showing chromosomal instability among 21 patients afflicted with thyroid cancer versus the comparable percentage among 6 control subjects. The data are shown in the table that follows.

Let us begin by stating the null hypothesis and the alternative hypothesis, selecting the appropriate test, and deciding on the level of significance for rejecting the null hypothesis.

Percentage of Cells with Chromosomal Instability

Patients		Controls	
X_1	X_1^2	X_2	X_2^2
13.9	193.21	4.0	16
10.9	118.81	1.0	1
9.3	86.49	5.0	25
5.2	27.04	7.0	49
14.0	196.00	3.0	9
2.0	4.00	3.0	9
3.2	10.24		
12.0	144.00		
13.0	169.00		
11.3	127.69		
12.2	148.84		
5.0	25.00		
2.0	4.00		
2.0	4.00		
17.1	292.41		
13.4	179.56		
6.0	36.00		
14.0	196.00		
18.0	324.00		
13.6	184.96		
5.4	29.16		

$$\sum X_1 = 203.5 \qquad \sum X_2 = 23.0$$

$$\sum X_1^2 = 2500.41 \qquad \sum X_2^2 = 109.00$$

Since we are testing a directional experimental hypothesis (chromosomal instability is higher among thyroid cancer patients than among normal controls), our null hypothesis is $\mu_1 \geq \mu_2$. In other words, the null hypothesis is that the percentage of chromosomal instability among thyroid cancer patient is equal to or less than the population from which the control subjects were drawn. If we reject H_0 we may infer that chromosomal instability is greater among thyroid cancer patients. Since this is a directional hypothesis, we shall apply a one-tailed test of significance. We set α at the .01 level. Since our data are ratio and total N is small, the Student t-ratio is the appropriate test statistic. Since df = 25, our critical value is defined by $t \geq 2.485$.

Let's look at the step-by-step calculation of t.

Step 1. Find $\sum X_1$ and $\sum X_2$. In the present problem they are, respectively, 203.5 and 23.

Step 2. Find the mean of each condition. They are: $\bar{X}_1 = 203.5/21 = 9.69$; $\bar{X}_2 = 23/6 = 3.83$.

Step 3. Find $\sum X_1^2$ and $\sum X_2^2 \cdot \sum X_1^2 = 2500.41$; $X_2^2 = 109$.

Step 4. Find SS_1 and SS_2:

$$SS_1 = 2500.41 - (203)^2/21 = 528.40$$
$$SS_2 = 109 - (23)^2/6 = 20.83$$

Step 5. Find the t-ratio:

$$t = \frac{9.69 - 3.83}{\sqrt{\left(\frac{528.4 + 20.83}{21 + 6 - 2}\right)\left(\frac{1}{21} + \frac{1}{6}\right)}}$$

$$= \frac{5.86}{\sqrt{(21.97)(27/126)}}$$

$$= 5.86/\sqrt{4.71} = 5.86/2.17 = 2.700 \quad df = 25$$

Since the obtained t-ratio is greater than the critical value of 2.485, we reject the null hypothesis and conclude that the two samples were drawn from different populations of chromosomal instabilities. Chromosomal instabilities are significantly greater among thyroid cancer patients than among normal controls. However, we must bear in mind that this study is not a true experiment since the subjects were not and could not be assigned at random to experimental conditions. Consequently, we cannot conclude that chromosomal instability was the *cause* of the thyroid cancer. Other possibilities exist such as: the cancer process is causing chromosomal instability; whatever caused the cancer may also have caused the chromosomal instability; the treatment received by the patients (radiation, chemotherapy) may have caused chromosomal instability.

On the basis of the t-test, we concluded that the samples were drawn from different populations of means, but an underlying assumption of the t-ratio is that the variances are homogeneous; that is, they are not significantly different. By using the F-distribution we could test this assumption. In the preceding study, the estimated variance ($\hat{s}^2$) of the patient group equals $528.40/20 = 26.42$, whereas, for the controls, $\hat{s}^2 = 20.83/5 = 4.17$.

To test for homogeneity of variances, the formula for F is:

$$F = \frac{\text{larger variance}}{\text{smaller variance}} = \frac{26.42}{4.17}$$

Substituting the estimated variances from the chromosomal study, we determine that F is 6.34. Turn to Table D_1 in the appendix to the text, which provides the critical lower and upper values of F required for significance at the .05 level. The columns represent the degrees of freedom for the numerator, and the rows represent the df for the denominator. Given the degrees of freedom for the numerator (the patient group) equals 20 and the df for the controls equals 5, we find two tabled values. Look at the bottom value: 6.33. Since the obtained F of 6.34 exceeds 6.33, we reject H_0 at the .05 level. Thus, the estimated variances are significantly different from one another. A significant F-ratio suggests the possibility that the experimental conditions may have produced a dual effect among the subjects. For example, some thyroid cancer patients may have had higher than expected chromosomal instabilities and others lower. A significant F in the test for homogeneity of variances should alert us to check for two effects of the experimental conditions on the sample subjects.

In generalizing from samples to populations, there is ample room for the researcher to make an error. As a result of the variability inherent in the data, a behavioral scientist can never be certain of his or her experimental conclusions.

This is why we must discuss significance tests and probability values. Even with the precautions and safeguards against error with which an investigator in the behavioral sciences is concerned, he or she can never prove the null hypothesis; it can only be accepted or rejected on the basis of a probability assessment. When making inferences about population parameters, the results are tentative at best.

Utilizing correlated samples is one way of minimizing error. If we know how much of a relationship exists between the criterion scores of matched samples, we can, in effect, statistically remove this source of error from our analysis of the data. In this section, we shall examine correlated samples, rather than samples drawn at random from a population.

When behavioral scientists set up an experiment, any single score they obtain on an individual may contain a component of any or all of three factors: (1) the subject's ability on the criterion task, (2) the effects of the experimental variable, and (3) the random error caused by a variety of reasons, for example, perhaps the subject is ill that day or the experimental procedures are highly unreliable. Of course, of these three factors, the investigator wishes to pinpoint the effects of the experimental or independent variable as precisely as possible. It is for this information about the experimental variable that the study is being performed.

As a consequence of wishing to determine with as much precision as possible the effects of the independent variable, the experimenter also must endeavor to minimize any effects of the subject's ability or of random error in the outcome of the experiment. Although random error cannot be entirely eliminated, most scientists make rigorous attempts to maintain consistent procedures in the experimental setting and eliminate distractions that contribute to error. In addition, through the use of the correlated samples, methods have been developed to quantify and, subsequently, remove the effects of the subject's ability or level of performance from the standard error. It is these procedures of statistically removing the effects of the subject's level of performance on the task to which we address ourselves in this section. Rather than reflecting on the individual differences among subjects, the results of an experiment should show the effects of the experimental variable. We shall now explore the methods devised to remove statistically from the error the influence of individual differences among subjects.

As we noted earlier in the chapter, the error term in the Student's t-ratio can be represented by the following formula:

$$s_{\bar{X}_1 - \bar{X}_2} = \sqrt{s_{\bar{X}_1}^2 + s_{\bar{X}_2}^2}$$

However, we intentionally failed to mention that this is the appropriate formula *only* when the samples are drawn at random from the population. If correlated samples are used and $r \neq 0$, then there is a more general formula for $s_{\bar{X}_1 - \bar{X}_2}$:

$$s_{\bar{X}_1 - \bar{X}_2} = \sqrt{s_{\bar{X}_1}^2 + s_{\bar{X}_2}^2 - 2r\, s_{\bar{X}_1 - \bar{X}_2}}$$

As you can see, the two formulas differ in that the latter has an additional term under the square root sign. When the samples are correlated, $r \neq 0$ and the term will have a value greater than zero. When $r = 0$, the quantity becomes zero and the term disappears from the formula.

There are many instances in which the samples are not selected randomly; that is, they are correlated.

In these cases, by applying the more general formula for $s_{\bar{X}_1 - \bar{X}_2}$, we can reduce the error associated with estimating the difference between means. In other words, by taking into account the correlation between the samples, we can produce a more sensitive test of the difference between means, a test that is more likely to lead to the rejection of the null hypothesis when H_0 is false. When we say that the test is "more sensitive," we are stating in effect that it is a more "powerful" test when H_0 is false. Power is defined as the probability of rejecting H_0 when it is actually false. The more powerful a test, the more likely we are to reject a false H_0. With correlated samples, we can gain a more powerful test of the difference between means. Of course, the greater the correlation between the samples is, the more powerful the test becomes when H_0 is false.

The drawback to correlated samples design is that we must calculate our degrees of freedom according to the formula

$$df = N - 1$$

where N = the number of pairs in the correlated samples. So, although we gain power when there is a correlation between samples, we lose degrees of freedom in our significance testing. Thus, a larger t-ratio is required for statistical significance.

There are many types of experimental situations in which correlated samples are appropriate. Do most students perform better in a relaxed classroom environment or in a more structured one? Is behavior modification an effective tool for improving the speech of individuals who stutter? Is there a higher degree of social adaptiveness at the age of 24 months than at 18 months? Each of these situations is suited to a correlated sample analysis.

There are two classifications of situations in which subjects are not selected at random from the population.

1. *Before-after design.* The same subjects are scored twice on some measure, once before the introduction of the experimental variable and once after it. As you can see, the second sample of individuals is identical to the first. To test the effectiveness of behavior modification as a therapy for stutterers, the same individuals would be scored on the extent of stuttering both before and after behavior modification treatment. Naturally, we would expect a correlation between the two dependent measures.

2. *Matched group design.* Although the two samples are composed of different individuals, members of both the experimental and the control groups are matched on some variable that is known to be correlated to the criterion or dependent variable. If we wanted to know the effects of a relaxed versus a structured classroom environment on student academic performance, we would want to match the students in the two groups on some variable that is correlated with academic performance. In much research, subjects are matched on IQ. This means that for each individual in Group A with an IQ of 115, there would be a member of Group B with approximately the same IQ. Since IQ is known to correlate with classroom performance, IQ would be a variable on which the subjects would be matched.

Now that we are aware that we shall find correlated samples in both before and after designs, let us attack a sample problem. Although $s_{\bar{X}_1 - \bar{X}_2}$ can be determined by correlating the samples, there exists a simpler way of finding $s_{\bar{X}_1 - \bar{X}_2}$ — the direct-difference method. Represented by the notation $s_{\bar{D}}$, the direct-difference method involves obtaining the differences between the pairs of matched subjects on the measure of the dependent variable. Once a difference score is found for each pair of subjects, the difference scores should be treated as if they were raw scores. This means that the difference scores, denoted by D, can be used in the t-ratio, the sum of squares, the formula for the standard deviation of a distribution of differences, and, finally, the standard error of the mean difference.

The formula for the t-ratio, assuming $\mu_D = 0$, is

$$t = \frac{\bar{D} - \mu_D}{s_{\bar{D}}} = \frac{\bar{D}}{s_{\bar{D}}}$$

where $\bar{D} = \sum D / N$.

We shall also need the formula for the sum of squares:

$$SS_D = \sum D^2 - (\sum D)^2 / N$$

With the formula for the sum of squares the standard deviation of the difference scores can be calculated with ease:

$$\hat{s}_D = \sqrt{\frac{SS_D}{N - 1}}$$

And the standard error of the difference formula using the direct-difference method becomes

$$s_{\bar{D}} = \sqrt{\frac{SS_D}{N(N - 1)}}$$

A team of psychologists attempting to teach chimpanzees to communicate with sign language is interested in ascertaining whether the chimps acquire the signing ability more readily when they live among other chimpanzees who are also learning sign language or when they are isolated from chimpanzees and their only contacts are the human instructors. One month the chimps are allowed to live with other chimpanzees, the next month they live only with their instructors. At the end of each month, they are scored on the acquisition of sign language. The following is a table of the raw data (hypothetical):

Subject	First month	Second month	Difference	
1	20	16	4	16
2	12	13	−1	1
3	6	10	−4	16
4	19	20	−1	1
5	11	13	−2	4

Let us follow the formal procedure for setting up a problem.

1. Null hypothesis: There is no difference in the means of the populations from which the before and after scores were drawn, that is, $\mu_D = 0$.

2. Alternative hypothesis: Since we have no idea in which direction the scores might be slanted (in favor of the first month or the second month), we shall postulate a nondirectional difference, $\mu_D \neq 0$.

Note: If we had a compelling theoretical reason to anticipate a better performance (higher scores) on the first-month round of scores, the difference scores would have been positive and H_1 would have been $\mu_D > 0$. On the other hand, if we had expected the second-month scores to be higher, our hypothesis would have been $\mu_D < 0$. In this case, the difference scores would have been negative.

Since H_1 is nondirectional, we shall conduct a two-tailed test of significance.

3. Statistical test: Student t-ratio.

4. Significance level: $\alpha = .01$.

5. Sampling distribution: Applying the degrees of freedom formula, we find that df $= N - 1 = 4$.

6. Critical region: $t_{.01} \leq -4.604$ or $t_{.01} \geq 4.604$.

Using the formulas we mentioned previously, you can find that the sum of squares is:

$$\begin{aligned} SS_D &= 38 - (-4)^2/5 \\ &= 34.8 \end{aligned}$$

Also,

$$\begin{aligned} s_{\bar{D}} &= \sqrt{34.8/5 \cdot 4} = \sqrt{1.74} \\ &= 1.32 \end{aligned}$$

Since $\bar{D} = .8$, we can substitute the appropriate values in the formula for t to obtain

$$t = 0.8/1.32 = .61$$

As you can quickly tell, t does not fall in the critical region. Therefore, we can accept H_0.

SELECTED EXERCISES

1. Explain what is meant by a difference that is statistically significant at any given level of significance.

2. In testing $H_0{:}\mu_1 = \mu_2$, if there is a real difference in the population means, that is, $\mu_1 \neq \mu_2$, will the obtained difference between the sample means always be statistically significant? Explain.

3. When would we use z as the statistic to test for the difference between means?

4. What are the assumptions underlying the use of the t-distributions?

5. What must be true about $s_{\bar{x}_1 - \bar{x}_2}$ in order that

$$\frac{(\bar{X}_1 - \bar{X}_2) - (\mu_1 - \mu_2)}{s_{\bar{x}_1 - \bar{x}_2}}$$

may be used as the test statistic in testing hypotheses?

6. How serious is it if $\sigma_1^2 \neq \sigma_2^2$? This is a violation of which assumption?

Discuss the implications of violating this assumption.

7. The manager of a school cafeteria at a large university believed that women consume more coffee daily than men. He asked a total of 20 randomly selected students on randomly selected days how many cups of coffee they had consumed during the previous 24-hour period. He obtained the following results. What did he conclude?

Men		Women	
5	1	3	7
3	3	5	1
10	1	3	2
3	8	6	8
1	3	5	
0			

8. An instructor hypothesizes that the students who earn C or better spend more time outside of class on the course work than students who receive a D or F. She collects the following data from two independent samples of students. What does she conclude?

No. of hours per week			
C or better		D or F	
8	5	6	3
4	3	2	2
4	2	1	1
2	1	0	5
1	3		

9. A mathematics instructor was interested in determining whether men students made better marks than women in mathematics. He randomly selected the final test cores of ten men and ten women. Is the difference between groups significant at the 5% level?

Men		Women	
32	48	38	62
35	52	43	57
37	46	38	49
44	51	39	61
47	59	46	58

10. Employing the data in the preceding problem, test for homogeneity of variance.

11. A home economist wished to determine whether there was any difference in the amount of fruit contained in 8-ounce cans produced by two competing companies. She randomly selected 50 cans from each company and analyzed the contents. The results were as follows:

Company I	Company II
$\bar{X}_1 = 8.3$	$\bar{X}_2 = 7.8$
$SS_1 = 480.5$	$SS_2 = 364.5$

What did she conclude?

12. For each sample in the preceding Selected Exercise, test the hypothesis $H_0:\mu_0 = 8.0$.

13. An investigator hypothesizes that the daily caloric intake of single women differs from that of married women. He selects two random samples, all between the ages of 25 and 30 and all approximately the same height and build. He obtains the following results. What does he conclude? (Data are in hundreds of calories.)

Single	Married
22	20
21	20
20	21
23	21
19	22
20	22
21	24

14. An investigator was interested in whether people who vacation at a certain resort on the American plan (meals included) tend to gain more weight on their vacation than people at the same resort on the European plan (no meals included). He selected two random samples and recorded the change in weight over a 2-week period. What did he conclude?

American plan		European plan	
+6	0	−1	+3
+9	+5	0	−5
0	−3	+4	+4
−2	+5	0	+1
+2	−1	+1	−1
+1	0	0	+6

15. An investigator believed that the life of batteries is affected by the way they are used. One sample was subjected to continuous use, the other, to frequent rest periods. He tested both samples of batteries with the same kind of light bulbs and recorded the total number of life hours. What did he conclude? Calculate ω^2.

Continuous		Rest periods	
20	18	20	20
21	19	20	26
19	20	25	22
17	18	23	24
20	19	24	25

16. What effect does matching have on the standard error of the difference between means? Why?

17. What are the degrees of freedom in the *t*-test for correlated samples? How does this differ from the degrees of freedom in the *t*-test for independent groups?

18. What is the disadvantage of matching samples on a variable that is not correlated with the criterion variable?

19. An investigator wishes to determine the effect of different degrees of motivation on the performance of a particular task. He employs seven sets of identical twins and randomly assigns them to the two groups. What does he conclude?

Group I	Group II
7	9
11	10
4	8
11	9
9	11
7	18
12	15

20. The manager of a large supermarket wanted to find out if people purchased products on the basis of brand name or on the basis of price. He set up two identical displays—one containing a well-known brand name product, the other containing the identical brand, selling at a lower price. He recorded the number of sales for each day of the week. What did he conclude?

	Brand name	Lower price
Monday	134	117
Tuesday	59	51
Wednesday	61	66
Thursday	77	83
Friday	129	125
Saturday	168	144

21. Two groups of college students, 50 in each group, were matched on grade point average. One group was taught by the lecture method with the entire group in one class. The other group was divided into four smaller discussion classes. At the end of the experimental period all students were given the same final examination. The correlation between grade point average and final test score was .62. The following results were obtained on the final test:

Lecture	Discussion
$\bar{X}_1 = 68.4$	$\bar{X}_2 = 71.8$
$s_1 = 6.3$	$s_2 = 7.0$

What is your conclusion?

22. An investigator believed that people who smoke tend to smoke more during periods of stress. She compared the number of cigarettes ordinarily smoked by a group of 15 randomly selected students with the number they smoked

during the 24 hours prior to final examinations. What did the investigator conclude?

Usual no.	Prior to final	Usual no.	Prior to final
8	7	36	32
15	10	18	21
22	28	16	18
11	12	15	14
17	19	17	22
10	9	9	9
6	7	11	16
31	39		

23. The owner of a small grocery store claimed that, on the average, his prices were the same as those of the large neighborhood supermarket. He compared the prices of ten randomly selected products. What did he conclude?

Grocery	Supermarket	Grocery	Supermarket
0.98	0.86	0.48	0.49
0.23	0.23	0.88	0.79
0.18	0.18	1.33	1.29
0.42	0.39	1.82	1.75
0.57	0.63	1.11	0.99

24. The dean of a large university believed that college seniors who intend to go to graduate school earn better grades than those who do not. He recorded senior year grade point averages for two groups of seniors, matched on the basis of their cumulative averages for the preceding three years. What did he conclude?

Graduate school candidates	Noncandidates	Graduate school candidates	Noncandidates
3.5	3.4	2.3	2.6
2.9	3.1	2.9	3.3
3.6	3.4	3.9	4.0
3.7	3.7	2.5	1.9
4.0	4.0	3.4	2.8
2.7	2.3	3.7	3.5
3.1	2.8		

Answers:

1. A difference in statistics is statistically significant if its occurrence is rare. Rare is defined in terms of probability. If the observed difference would have occurred by chance 5 times or less in 100, the level of significance is .05. If 1 time in 100 or less, the level of significance is .01. The level of significance is selected prior to collecting and analyzing the data.

2. No. Sample means do not precisely reflect their corresponding population values. Sampling error can lead to differences in means that are less than (or greater than) the differences that exist in the population.

3. When we know that the population distribution is normal and we know the true value of the population variances. Also, if very large samples are

selected from normally distributed populations, the difference between the critical values of z and t become negligible.

4. The scores are interval/ratio and the populations are normal with homogeneous variances.

5. The estimated standard error of the differences between means is based on the unbiased estimate of the population variance.

6. The importance of the violation of the assumption of homogeneity of variances is hotly debated. Some argue that the t-test is robust and will survive modest violations of the assumption of homogeneity. Others argue that the value of the t-ratio is only as good as the extent to which it fulfills its underlying assumptions. In practice, the true form of the underlying distributions is rarely known with any precision. However, what is clear is that the finding of heterogeneity should alert us to the possibility that more than one effect is being exerted by the experimental variable.

7.

Men	Women
$\sum X_1 = 38$	$\sum X_2 = 40$
$\sum X_1^2 = 228$	$\sum X_2^2 = 222$
$SS_1 = 96.73$	$SS_2 = 44.22$

$$t = \frac{3.45 - 4.44}{\sqrt{\dfrac{96.73 + 44.22}{18}\left(\dfrac{1}{11} + \dfrac{1}{9}\right)}} = \frac{-0.99}{1.26}$$

$t = 0.786, \ df = 18$

8.

C or better	D or F
$\sum X_1 = 33$	$\sum X_2 = 20$
$\sum X_1^2 = 149$	$\sum X_2^2 = 80$
$SS_1 = 40.1$	$SS_2 = 30$

$$t = \frac{3.3 - 2.5}{0.99}$$

$t = 0.808, \ df = 16$

9.

Men	Women
$\sum X_1 = 451$	$\sum X_2 = 491$
$\sum X_1^2 = 20,969$	$\sum X_2^2 = 24,953$
$SS_1 = 628.9$	$SS_2 = 844.9$

$$t = \frac{45.1 - 49.1}{4.05}$$

$$t = .988, \ df = 18$$

10. $s_1^2 = 69.88$, $s_2^2 = 93.88$, $F = 1.34$, $df = 9/9$

11. $t = 0.851$, $df = 98$

12. I:　$t = .677$,　$df = 49$
    II:　$t = .519$,　$df = 49$

13.

Single	Married
$\sum X_1 = 146$	$\sum X_2 = 150$
$\sum X_1^2 = 3056$	$\sum X_2^2 = 3226$
$SS_1 = 10.86$	$SS_2 = 11.71$

$$t = \frac{20.86 - 21.43}{.73}$$

$$t = .78, \ df = 12$$

14.

American	European
$\sum X_1 = 22$	$\sum X_2 = 12$
$\sum X_1^2 = 186$	$\sum X_2^2 = 106$
$SS_1 = 145.67$	$SS_2 = 94$

$$t = \frac{1.83 - 1.00}{1.35}$$

$$t = .615, \ df = 22$$

15.

Continuous	Rest periods
$\sum X_1 = 191$	$\sum X_2 = 229$
$\sum X_1^2 = 3661$	$\sum X_2^2 = 5291$
$SS_1 = 12.9$	$SS_2 = 46.9$

$$t = \frac{19.1 - 22.9}{.82}$$

$$t = 4.634, \ df = 18$$

$$\omega^2 = .51$$

16. It has no effect if the correlation between the matching variable and the dependent variable is zero. However, if there is a correlation between the two variables, the magnitude of the standard error is decreased. The larger the correlation is, the greater is the decrease.

17. It is the number of pairs minus 1. Thus, if $N_1 = N_2 = 15$, df $= 15 - 1 = 14$. For independent samples, df $= N_1 + N_2 - 2$. If $N_1 = N_2 = 15$, df $= 15 + 15 - 2 = 28$. For a given number of observations in each group, the number of degrees of freedom for matched pairs is 1/2 the number of degrees of freedom for independent samples.

18. There is a loss of degrees of freedom necessitating a larger t-ratio for statistical significance. If the correlation is zero, there is no compensatory reduction in the size of the error term.

19. $\sum D = -19$ $\quad \sum D^2 = 159$
    $\bar{D} = -2.71$ $\quad s_{\bar{D}} = 1.60$
    $\quad\quad t = -1.70,$ df $= 6$

20. $\sum D = 42$ $\quad \sum D^2 = 1006$
    $\bar{D} = 7$ $\quad s_{\bar{D}} = 4.87$
    $\quad\quad t = 1.437,$ df $= 5$

21. $\bar{D} = -3.4$ $\quad$ df $= 49$
    $s^2_{\bar{X}_1} = .66$ $\quad s^2_{\bar{X}_2} = 1.0$
    $s_{\bar{X}_1} = .81$ $\quad s_{\bar{X}_2} = 1.0$
    $s_{\bar{D}} = .81$ $\quad t = -4.198$
    reject H_0

22. $\sum D = -21$ $\quad \sum D^2 = 213$
    $\bar{D} = -1.4$ $\quad s_{\bar{D}} = .935$
    $\quad\quad t = -1.50,$ df $= 14$

23. $\sum D = .4$ $\quad \sum D^2 = .048$
    $\bar{D} = .04$ $\quad s_{\bar{D}} = .019$
    $\quad\quad t = 2.11,$ df $= 9$

24. $\sum D = 1.4$ $\quad \sum D^2 = 1.36$
    $\bar{D} = .108$ $\quad s_{\bar{D}} = .088$
    $\quad\quad t = 1.23,$ df $= 12$

SELF-QUIZ: TRUE-FALSE

Circle T or F.

T   F   1. The population standard deviations must be known in order to employ the t-ratio.

T   F   2. When we obtain a negative t-ratio, we have made an error in calculation.

T   F   3. A significant difference in variances increases the likelihood of rejecting $H_0: \mu_1 = \mu_2$.

T   F   4. The mean of the sampling distribution of the difference between means will always be zero.

T   F   5. The sampling distribution of the statistic

$$t = \frac{(\bar{X}_1 - \bar{X}_2) - (\mu_1 - \mu_2)}{s_{\bar{X}_1 - \bar{X}_2}}$$

is, by definition, normal.

T   F   6. One of the assumptions underlying the use of the t-distribution is that the samples are drawn from populations whose means are equal.

T   F   7. The mean of the sampling distribution of the difference between means equals μ.

T   F   8. If the value of t does not fall within the critical region, we may conclude that $H_0: (\mu_1 - \mu_2) = 0$ is true.

T   F   9. When we accept H_0, we conclude that both samples come from the same population.

T F 10. In the two-sample case, the null hypothesis states the expected value of $(\bar{X}_1 - \bar{X}_2)$.

T F 11. $(\mu_1 - \mu_2)$ is based on an unbiased estimate of $(\bar{X}_1 - \bar{X}_2)$.

T F 12. $H_1{:}(\mu_1 \neq \mu_2)$ is a nondirectional alternative hypothesis.

T F 13. A necessary assumption underlying the use of the z-statistic is that the N's are large.

T F 14. A negative t-ratio means that the samples were drawn from populations whose means differ by less than the value stated in the null hypothesis.

T F 15. The sampling distribution of the z-statistic is normal with df $= N_1 + N_2 - 2$.

T F 16. When the degrees of freedom are small, larger t-ratios are required for significance.

T F 17. Most inferential analyses in the behavioral sciences involve a single sample.

T F 18. With equal N's the estimated standard error of the difference between means can be expressed as

$$\sqrt{\frac{SS_1 + SS_2}{N(N-1)}}$$

T F 19. A matched group design reduces the likelihood of a Type I error.

T F 20. The Student t-ratio for independent groups yields the same probability values as Student's t-ratio for correlated samples when the correlation between paired observations is 0.

T F 21. Employing correlated samples usually improves our ability to estimate the effects of the experimental variable on the dependent measures.

T F 22. The score of any subject on the criterion variable represents the effects of the experimental variable only.

T F 23. We employ correlated samples to "statistically remove" the effects of error contributed by individual differences.

T F 24. When scores are paired at random, the correlation between the two samples will average zero.

T F 25. When we assign subjects to experimental conditions at random, we are assured that the experimental groups are "equivalent" in initial ability.

T F 26. As r approaches zero, the advantage of employing correlated samples becomes progressively smaller.

T F 27. The advantage of matching is that a major source of variation is identified and removed from error.

T F 28. When we employ the direct-difference method, it is not necessary to calculate the correlation between samples.

T F 29. A negative value of $s_{\bar{D}}$ indicates that $\mu_{\bar{D}} < 0$.

T F 30. Employing $\alpha = .05$, two-tailed test, we obtain $A = .275$, in a test comparing ten pairs of subjects. We reject H_0.

T   F   31. If the obtained value of t for correlated samples does not fall within the critical region, then the value for t for independent samples obtained for the same data cannot be statistically significant.

T   F   32. Adding a constant to all scores in a correlated sample design will not change the difference between means or the standard error of the difference.

T   F   33. Random error and the subject's ability on the criterion task are factors in the variability of criterion scores.

T   F   34. The matched design is composed of two samples of the same individuals, whereas the before-after design includes two unique samples.

Answers: (1) F; (2) F; (3) F; (4) F; (5) F; (6) F; (7) F; (8) F; (9) F (10) F; (11) F; (12) T; (13) F; (14) F; (15) F; (16) T; (17) F; (18) T; (19) F; (20) F; (21) T; (22) F; (23) T; (24) T; (25) F; (26) T; (27) T; (28) T; (29) F; (30) T; (31) F; (32) T; (33) T; (34) F.

SELF-TEST: MULTIPLE-CHOICE

1. Which of the following is *not* an acceptable statement of a null hypothesis?

   a) The proportion of Democrats among registered voters is .51.
   b) The mean IQ of the population is 106.
   c) The difference between the two sample means is 0.
   d) The two samples were drawn from the same population.
   e) The two samples were drawn from two populations with a mean difference of 2.00.

2. An experimenter randomly assigns nine people to an experimental group and another nine people to a control group, thus assuring the independence of the two groups. How many degrees of freedom are there in the experimenter's t-test?

   a) 18   b) 8   c) 9   d) 17   e) 16

3. Which of the following is an assumption required for the correct use and interpretation of the standard error of the difference between means?

   a) that the population N is 100 or more
   b) that the samples have an N of less than 30
   c) that the obtained samples are derived from two different populations, both of which are normally distributed
   d) that the differences between pairs of sample means from the same population are normally distributed
   e) that the population standard deviations are known

4. Which of the following values must one know in order to test the null hypothesis based on the difference between means?

   a) the mean of the population from which the samples are drawn
   b) the population N
   c) the correlation of the sample means
   d) the variability of the sample means
   e) the standard deviation of the population

5. The mean of the sampling distribution of the difference between pairs of sample means drawn at random from the same normally distributed population is:

   a) 0.00

b) either $z = +1.00$ or $z = -1.00$
c) greater than $z = +2.58$ or less than $z = -2.58$
d) unknown because the difference depends directly on the σ of the population
e) the standard error of the difference between means

6. Which of the following is *not* an assumption underlying the *t*-test?

a) normality
b) random sampling
c) homogeneity of variance
d) unbiased estimate of the population variance
e) discrete scales underlying measurement

7. If we say that there is a real difference between two groups of data, but in reality there is no difference, we have committed what kind of error?

a) Type I error   b) Type II error   c) Type III error   d) beta error
e) two-tailed error

8. The estimated standard error of the difference between means for independent groups equals:

a) $\sqrt{\dfrac{SS_1 + SS_2}{N_1 + N_2 - 2} \left(\dfrac{1}{N_1} + \dfrac{1}{N_2} \right)}$ b) $\sqrt{\dfrac{s_{\bar{X}_1}^2 + s_{\bar{X}_2}^2}{N - 1}}$

c) $\sqrt{\dfrac{s_{\bar{X}_1}^2 + s_{\bar{X}_2}^2}{N}}$ d) $\dfrac{\sigma_{\bar{X}_1 - \bar{X}_2}}{\sqrt{N}}$ e) $\dfrac{\sigma_{\bar{X}_1 - \bar{X}_2}}{\sqrt{N - 1}}$

9. The mean of the sampling distribution of the mean difference between pairs of sample means drawn from the *same* population is:

a) 0   b) 1   c) a function of the size of the samples
d) a function of $\sigma_{\bar{X}_1 - \bar{X}_2}$ e) never less than $\sigma_{\bar{X}_1 - \bar{X}_2}$

10. Which of the following could *not* serve as a null hypothesis in the two-sample case?

a) $\mu_1 - \mu_2 = 0$ b) $\mu_1 - \mu_2 = 5$ c) $\mu_1 - \mu_2 = -5$ d) $\bar{X}_1 - \bar{X}_2 = 0$
e) all of the above are acceptable null hypotheses

11. Which of the following is *not* an assumption underlying the use of the *t*-distribution?

a) the sampling distribution of the difference between means is normally distributed
b) $s_{\bar{X}_1 - \bar{X}_2}$ is based on the unbiased estimate of the population variance
c) both samples are drawn from populations whose variances are equal
d) the standard deviation of each population is known
e) none of the above

12. Which of the following is a true statement?

a) the *t*-distributions are symmetrical about a mean of 1.00
b) the *t*-ratio is employed only when the standard deviation of each population is known
c) we accept H_0 when we obtain a negative t-ratio
d) we employ the *F*-ratio to test for homogeneity of variance
e) none of the above is a true statement

Questions 13 through 16 refer to the following data:

$$\bar{X}_1 = 100 \quad \bar{X}_2 = 90 \quad N_1 = 10 \quad N_2 = 10 \quad SS_1 = 1440 \quad SS_2 = 1440$$

13. $s_{\bar{X}_1 - \bar{X}_2}$ equals:

   a) 1.20   b) 1.33   c) 4.00   d) 5.66   e) 10.00

14. The obtained t-ratio for testing $H_0 : \mu_1 = \mu_2$ equals:

   a) 1.000   b) 1.767   c) 2.500   d) 7.519   e) 8.333

15. The df is:

   a) 8   b) 9   c) 10   d) 18   e) 19

16. Employing $\alpha = .05$, *one-tailed test,* you would conclude that:

   a) the obtained t-ratio does not fall within the critical region
   b) the obtained $t < t_{.05}$ at $\alpha = .05$, one-tailed test
   c) there was no significant difference between the means
   d) the null hypothesis was accepted
   e) the null hypothesis was rejected

17. If we randomly draw pairs of samples of a fixed N from a population in which $\mu = 78$, $\sigma = 10$, and obtain a frequency distribution of differences between each pair, we would expect $\mu_{\bar{X}_1 - \bar{X}_2}$ to equal:

   a) 78.00   b) 10.00   c) 0.00   d) 7.80
   e) cannot answer question without knowing N

18. If we randomly select pairs of samples of a fixed N from two populations in which $\mu_1 = 72$, $\sigma_1 = 10$, and $\mu_2 = 75$, $\sigma_2 = 10$, and calculate the difference between these pairs of sample means, the resulting distribution would have a mean $(\mu_{\bar{X}_1 - \bar{X}_2})$ equal to:

   a) 0.00   b) $-.3$ c) 3.00   d) -3.00
   e) cannot answer without knowing N

   Employing the following excerpt of critical values of t (Table C of Appendix D in the text), answer Multiple-Choice Problems 19 through 24.

Table C   Critical Values of t

df	\.10	\.05	\.025	\.01	\.005	\.0005
	\.20	\.10	\.05	\.02	\.01	\.001
6	1.440	1.943	2.447	3.143	3.707	5.959
7	1.415	1.895	2.365	2.998	3.499	5.405
8	1.397	1.860	2.306	2.896	3.355	5.041
9	1.383	1.833	2.262	2.821	3.250	4.781
10	1.372	1.812	2.228	2.764	3.169	4.587
11	1.363	1.796	2.201	2.718	3.106	4.437
12	1.356	1.782	2.179	2.681	3.055	4.318

The column groups are headed: **Level of significance for one-tailed test** (.10, .05, .025, .01, .005, .0005) and **Level of significance for two-tailed test** (.20, .10, .05, .02, .01, .001).

19. In an experiment comparing two groups on a criterion variable in which H_0 is $\mu_1 = \mu_2$, $\alpha = .05$, two-tailed test, $N_1 = 5$, $N_2 = 7$, a t-ratio of 2.179 is obtained. We should:

   a) reject H_0

b) fail to reject H_0
c) assert that there is no difference in the population means from which the samples were drawn
d) conclude that $\bar{X}_1$ is significantly greater than $\bar{X}_2$
e) conclude that we must know the sample means in order to answer the question

20. Given $\bar{X}_1 = 20$, $N_1 = 6$, $\bar{X}_2 = 15$, $N_2 = 5$, $s_{\bar{X}_1 - \bar{X}_2} = 2.5$, $\alpha = .05$, one-tailed test, our conclusion would be:

a) since df = 11, accept H_0 b) since df = 9, reject H_0
c) since df = 9, accept H_0 d) since df = 11, reject H_0
e) none of the above

21. Given $\bar{X}_1 = 20$, $N_1 = 6$, $\bar{X}_2 = 15$, $N_2 = 5$, $s_{\bar{X}_1 - \bar{X}_2} = 2.5$, $\alpha = .05$, one-tailed test, our conclusion would be:

a) since df = 11, accept H_0
b) since df = 9, reject H_0
c) since df = 9, accept H_0
d) since df = 11, reject H_0
e) none of the above

22. Given $N_1 = 7$, $N_2 = 3$, $s_{\bar{X}_1 - \bar{X}_2} = 3$, $\alpha = .01$, two-tailed test, we reject H_0 that $\mu_1 = \mu_2$. The difference in sample means is:

a) 8.688 or greater   b) less than 9.507   c) 9.507 or greater
d) less than 10.065   e) 10.065 or greater

23. Given $N_1 = 7$, $N_2 = 3$, $s_{\bar{X}_1 - \bar{X}_2} = 3$, $\alpha = .01$, two-tailed test, we fail to reject H_0 that $\mu_1 = \mu_2$. The difference in sample means is:

a) 8.688 or greater   b) less than 9.507   c) 9.507 or greater
d) less than 10.065   e) 10.065 or greater

24. Employing a two-group experimental design, a researcher tests $H_0 : \mu_1 = \mu_2$. He obtains a t-ratio of 2.316 and rejects H_0 at $\alpha = .05$, two-tailed test. In this experiment, $N_1 + N_2$ equals:

a) less than 8   b) 8 or greater   c) 10 or greater   d) less than 10
e) 7 or greater

25. Variations in behavior are due to variations in experimental conditions. This statement is:

a) proved when differences are statistically significant
b) proved when the null hypothesis is tenable
c) inferred when degree of correlation exceeds 5%
d) inferred when the null hypothesis is rejected
e) the null hypothesis

26. Given only the following data: A t-test between two correlated groups, Group A (control group) and Group B (test group), was run to test the effect of a certain drug. The degrees of freedom were 9, it was a two-tailed test, the critical value was 2.262, the level of significance was .05, and the computed value of t was 2.560. Which of the following statements is a logical conclusion from the above data?

a) the drug had no effect
b) Group A is superior to Group B
c) Group B is superior to Group A
d) there is a significant difference between Groups A and B
e) there is a difference between Groups A and B but it is not significant

27. When the data are so sampled that there is a correlation between the two dependent measures, using the standard error of a difference for independent measures tends to:

    a) decrease the power of the test
    b) overestimate the standard error of a difference
    c) increase the probability of a Type II error
    d) underestimate the value of t
    e) all of the above

28. The value $2rs_{\bar{x}_1-\bar{x}_2}$ usually operates to increase t by:

    a) increasing the correlation   b) reducing $s_{\bar{x}_1-\bar{x}_2}$ c) reducing $\bar{X}_2 - \bar{X}_1$
    d) increasing N e) increasing $s_{\bar{x}_1-\bar{x}_2}$

29. An experimenter makes a deduction from a specific theory that assumes in advance that one group will perform better than another. He employs matched groups in his experiment; $N = 25$ in each group. In order to reject the null hypothesis at the .05 level, t must exceed a value of:

    a) 2.064   b) 2.060   c) 1.711   d) 1.708   e) 1.714

30. In a given experiment there are two groups of five persons each. The individuals in one group are correlated with those in the other. What are the degrees of freedom for a t-test with this experiment?

    a) 4   b) 9   c) 8   d) 5   e) 3

31. An advantage of a matched group design is that:

    a) it ensures that the experimental groups are "equivalent" in initial ability
    b) it increases the likelihood of a Type I error
    c) it is easier to administer experimental treatments to matched subjects
    d) all of the above
    e) none of the above

32. An advantage of a matched group design is that:

    a) by employing matched pairs we reduce our degrees of freedom
    b) the higher the correlation between groups is, the more sensitive is the statistical test
    c) the higher the correlation is, the lower is the t-ratio required for significance
    d) all of the above
    e) none of the above

33. A disadvantage of a matched group design is that:

    a) the higher the correlation is, the lower is the error term
    b) there is an increased likelihood of a Type II error
    c) a smaller difference between means may be required for significance
    d) there is a loss of degrees of freedom
    e) all of the above

34. In a before-after design it is presumed that the experimental groups are correlated because:

    a) subjects have been carefully matched on a variable known to be correlated with the criterion variable

b) subjects have been assigned to experimental conditions at random
c) the assumption is necessary in order to permit the use of an error term reflecting correlated measures
d) it is assumed that each individual will remain relatively consistent in performance at different times
e) none of the above

35. Given: $\Sigma D = 20$, $\Sigma D^2 = 120$, $N = 5$, then $\bar{D}$ equals:

a) 24   b) 6.00   c) 4.00   d) .25   e) none of the above

36. Given: $\Sigma D = 20$, $\Sigma D^2 = 120$, $N = 5$, then $s_{\bar{D}}$ equals:

a) $\sqrt{40}$ b) $\sqrt{24}$ c) $\sqrt{5}$ d) $\sqrt{2}$ e) $\sqrt{10}$

37. Given: $\Sigma D = 24$, $SS_D = 120$, $N = 6$, the resulting t-ratio is:

a) 12.00   b) 2.00   c) .90   d) .54   e. 1.20

38. Given $\Sigma D = 20$, $\Sigma D^2 = 120$, $N = 5$, the Student t equals:

a) 60   b) .283   c) 5.350   d) 2.829   e) 2.309

39. With correlated samples, employing the formula for $s_{\bar{X}_1 - \bar{X}_2}$ based on matched groups usually:

a) yields a smaller error term
b) provides a more sensitive test of the differences
c) is more likely to lead to the rejection of a false H_0
d) is associated with a loss of df
e) all of the above

40. When subjects have been matched on a variable *uncorrelated* with the criterion variable, using $s_{\bar{X}_1 - \bar{X}_2}$ for matched groups:

a) results in smaller t-ratio
b) compensates for the loss of df
c) is associated with df $= N_1 + N_2 - 2$
d) greatly reduces the magnitude of $s_{\bar{X}_1 - \bar{X}_2}$
e) none of the above

41. The reactions of 12 subjects are compared on a psychomotor task before and after the injection of a drug. Employing $\alpha = .01$, two-tailed test, the t required for significance must be greater than or equal to:

a) 2.718   b) 2.681   c. 3.106   d) 4.587   e. 3.055

42. The most general formula for the standard error of the difference is:

a) $\sqrt{s_{\bar{X}_1}^2 + s_{\bar{X}_2}^2}$ b) $\sqrt{s_{\bar{X}_1}^2 + s_{\bar{X}_2}^2 - 2r\, s_{\bar{X}_1}^2 + s_{\bar{X}_2}^2}$

c) $\sqrt{\dfrac{SS_D}{N(N-1)}}$ d) $\sqrt{\dfrac{\Sigma D^2}{(\Sigma D)^2}}$ e) the direct-difference method

43. In an experiment employing a matched group design with ten pairs of subjects, $\Sigma D^2 = 112$, $\Sigma D = 20$. Employing $\alpha = .05$, one-tailed test, the obtained t:

a) will lead to acceptance of H_0
b) indicates a significant difference between the two conditions
c) is less than the value of t required for significance
d) all of the above
e) none of the above

44. Which of the following procedures in performing an experiment is most effective in eliminating the variation caused by chance factors?

    a) administer experimental procedures to small but independent samples
    b) observe the behavior of the same group of individuals under variations in experimental conditions
    c) select large samples that are known to be highly heterogeneous in makeup
    d) select the individuals to be observed by using a table of random numbers
    e) select a more stringent significance level

45. Which of the following procedures is necessary when testing the significance of the difference between means of two sets of measures obtained from a single group of individuals?

    a) use only one set of measures and obtain a new set of data from a second group of individuals who are independent of the first group
    b) assume that the two sets of measures are independent and derive the t-ratio in the standard way
    c) transform both sets of measures into z-scores and divide the difference of the mean z-score by 2.58
    d) use a standard error of the difference between means that takes into account the degree of correlation
    e) it is not possible to perform a test of significance on a single group of individuals

46. In which of the following instances is the standard error of the difference between means likely to be the smallest?

    a) when the scores in the two samples are paired and have positive correlations
    b) when scores are obtained from an experimental and a control group that are unrelated
    c) when the null hypothesis is rejected
    d) when the null hypothesis is accepted
    e) when the size of the samples is small

47. The number of correct discriminations on a 50-trial series is tabulated for 37 animals before and after brain surgery. What would be the degrees of freedom in testing the null hypothesis?

    a) 6   b) 49   c) 36   d) 72   e) 98

Answers: (1) c; (2) e; (3) d; (4) d; (5) a; (6) e; (7) a; (8) a; (9) a; (10) d; (11) d; (12) d; (13) d; (14) b; (15) d; (16) e; (17) c; (18) d; (19) b; (20) b; (21) c; (22) e; (23) d; (24) c; (25) d; (26) d; (27) e; (28) b; (29) c; (30) a; (31) a; (32) b; (33) d; (34) d; (35) c; (36) d; (37) b; (38) d; (39) e; (40) e; (41) c; (42) b; (43) a; (44) b; (45) d; (46) a; (47) c.

14

An Introduction to the Analysis of Variance

BEHAVIORAL OBJECTIVES

Conceptual Objectives

1. What is the purpose of the one-way analysis of variance (ANOVA)? When is it applicable?

2. Define and distinguish between the within-group variance and the between-group variance. What is their relationship to the magnitude of the F-ratio?

3. Explain how the total sum of squares, the between-group sum of squares, and the within-group sum of squares are derived. What is the relationship among them?

4. How are the variance estimates, the degrees of freedom, and the F-ratio obtained? Describe exactly what the null hypothesis predicts concerning the variance estimates and the F-ratio.

5. In relation to the F-ratio, what are the functions of planned and unplanned comparisons? When is Tukey's HSD test applicable?

6. What is the underlying assumption of homogeneity of variance?

7. How does a one-way correlated ANOVA differ from the one-way independent-samples ANOVA?

8. State the possible advantages and disadvantages of correlated-samples ANOVA from an independent-samples ANOVA.

Procedural Objectives

1. Given sample data, conduct an analysis of variance. Be prepared to fill in a summary table that includes the degrees of freedom, sum of squares, variance estimates, F-ratio, and level of significance. Familiarize yourself with the use of Table D in the appendix to the text in making decisions about the null hypothesis.

2. Conduct Tukey's HSD test of comparisons using Table O in the appendix to the text to determine the critical values.

3. Know how to conduct a one-way ANOVA correlated-samples design.

CHAPTER REVIEW

Now that we have discussed the two-sample comparison, we are ready to take the next step: the multigroup comparison. Just as the t-ratio is used with the two-sample case, the analysis of variance is the test for the multigroup comparison. As you continue your study of statistics or the behavioral sciences, you will discover that rarely do you compare two groups on a single variable. As a researcher, you will be more likely to find yourself involved with comparisons involving several levels of a single variable or even two or more variables. More complex experimental situations such as these require the use of the analysis of variance. In this discussion, we shall focus our attention on the multigroup case with a single independent variable.

As you have no doubt surmised, the analysis of variance is a statistical technique that enables the investigator to assess differences among many groups.

The sum of squares concept plays an invaluable part in the analysis of variance. As its name suggests, the analysis of variance is a method of comparing group differences on the basis of different variance estimates. To do so, we make use of the sum of squares concept.

In the single variable analysis of variance, the total sum of squares is divided into two components: within-group sum of squares (SS_W) and between-group sum of squares (SS_{bet}). Since the analysis of variance is built around the foundation of within-group and between-group sums of squares, you should be sure to master these concepts.

Stated simply enough, the within-group sum of squares is the combined sum of the squares that is obtained within each group. Denoted by SS_W, the within-group sum of squares is determined by adding the sum of the squares from each of the groups. With three groups, the within-group sum of squares formula is

$$SS_W = SS_1 + SS_2 + SS_3$$

To find the between-group sum of squares, we may use the following formula:

$$SS_{bet} = \Sigma N_i(\bar{X}_i - \bar{X}_{tot})^2$$

where N_i is the number in the ith group and $\bar{X}i$ is the mean of the ith group. If there are three groups in the experimental design, there will be three terms on the right-hand side of the equation in the formula for the between-group sum of squares.

As we have already mentioned, the within-group and between-group sums of squares are components of the total sum of squares. When added together, the values of the within-group and between-group sums of squares should equal that of the total sum of squares. Stated symbolically,

$$SS_{tot} = SS_w + SS_{bet}$$

The variance estimates of the between-group and within-group sums of squares are found merely by dividing each quantity by the appropriate number of degrees of freedom. For the between-group sum of squares, the degrees of freedom equal the number of cells minus one. The formula is

$$df_{bet} = k - 1$$

where k equals the number of cells.

For the within-groups, the degrees of freedom equal the total N of the combined groups minus the number of cells. The formula

$$df_W = N - k$$

can be used to arrive at the number of degrees of freedom for the within-groups. With a total N of 50 and 4 groups, $df_W = 46$.

For significance testing in the analysis of variance, we must use the F-ratio. As you know, the F-ratio is a ratio of variances. In the analysis of variance, the F-ratio is defined as the between-group variance estimate divided by the within-group variance estimate.

In symbols, this means

$$F = \hat{s}^2_{bet}/\hat{s}^2_W$$

For both these variance estimates, we merely divide the appropriate sum of squares by the corresponding degrees of freedom.

Using the information that has been presented thus far, let us work a three-group analysis of variance problem. In order to detect any significant differences in driving standards, three neighboring states randomly selected a group of ten drivers from each state. A behind-the-wheel examination was administered to all drivers. On the basis of these scores, representatives of the states wish to ascertain whether or not any significant differences in driving standards exist among the participating states. To make this decision, they will use the following information and show the step-by-step analysis.

| Driver | State A | | State B | | State C | |
	X_1	X_1^2	X_2	X_2^2	X_3	X_3^2
1	60	3,600	93	8,649	81	6,561
2	80	6,400	82	6,724	76	4,900
3	50	2,800	69	4,761	64	4,096
4	75	5,625	55	3,025	52	2,704
5	91	8,281	78	6,084	76	5,776
6	96	9,216	85	7,225	73	5,329
7	42	1,764	81	6,561	70	4,900
8	73	5,329	72	5,184	89	7,921
9	79	6,241	94	8,836	92	8,464
10	66	4,356	97	9,409	46	2,116
Sums	712	53,312	806	66,458	713	52,767

1. Find the sum of the scores of each group. In the accompanying example, $\Sigma X_1 = 712$, $\Sigma X_2 = 806$, and $\Sigma X_3 = 713$.

2. Add together each group sum to obtain the sum for the total (ΣX_{tot}).

$$\Sigma X_{tot} = 712 + 806 + 713 = 2231$$

3. Square each score and sum.

$$\Sigma X_{tot}^2 = 172,537$$

4. Find $SS_{tot} = \Sigma X_{tot}^2 - (\Sigma X_{tot})^2/N = 172,537 - 165,912.03 = 6,624.97$.

5. Find $SS_{bet} = (\Sigma X_1)^2/N_1 + (\Sigma X_2)^2/N_2 + (\Sigma X_3)^2/N_3 - \Sigma X_{tot}^2/N$.

However, since the N for each group is the same, we can square the sums and add them together before dividing by $N_k = 10$. Thus,

$$SS_{bet} = (712^2 + 806^2 + 713^2)/10 - 165,912.03 = 166,494.9 - 165,912.03$$
$$= 582.87$$

6. Find SS_W by subtraction. Thus,

$$SS_W = SS_{tot} - SS_{bet} = 6624.97 - 582.87 = 6042.1$$

7. Find the degrees of freedom for total, between, and within: $df_{tot} = N - 1 = 29$; $df_{bet} = 3 - 1 = 2$, $df_W = (k - 1)(N_k - 1) = 3(10 - 1) = 27$ (alternatively, $df_W = df_{tot} - df_{bet} = 29 - 2 = 27$).

8. Place the various sums of squares and degrees of freedom in the summary table.

Summary Table

Variable	Sum of squares	df	Estimated variance	F-ratio
Treatment	582.87	2	291.43	1.302
Within	6042.10	27	223.78	
Total	6624.97	29		

9. Divide the treatment sum of squares (SS_{bet}) by its degrees of freedom to obtain the estimated treatment variance. Thus, $582.87/2 = 291.43$. Enter this value into the summary table.

10. Divide the within sum of squares by its degrees of freedom to obtain the within variance estimate: $6042.10/27 = 223.78$. Enter this value into the summary table.

11. Divide the treatment variance estimate to obtain the F-ratio: $291.43/223.78 = 1.302$.

In the analysis of variance, the null hypothesis states that the two independent variance estimates may be regarded as estimates of the same population value. Symbolically speaking,

$$H_0: \mu_1 = \mu_2 = \mu_3 = \cdots = \mu_K$$

Since we have three groups in our example: $H_0: \mu_1 = \mu_2 = \mu_3$. If, when we consult Table D in the appendix to the text, we find we have a significant F-value, we then reject H_0.

Is our F significant at the .05 level? With 2 and 27 as our degrees of freedom, a value of 3.35 or greater is necessary for significance. (This critical value of F is obtained from Table D.) Since our F-value does not exceed the tabled value, we do not reject H_0. In other words, in the absence of evidence to the contrary, we act as if there is no difference among the driving standards of the three states.

Since our F-ratio did not prove to be significant, we conclude our statistical analyses at this point. If we had found significance, we would have tested specific hypotheses about the population parameters. Hypotheses about the population parameters can be separated into two general classes: (1) a priori or planned comparisons and (2) a posteriori comparisons. The Latin roots of these two terms should clue you as to the definitions of the categories. A priori comparisons are planned prior to the outset of the investigation, while a posteriori are not planned in advance but, rather, occur later (posterior) in the investigation.

In a multigroup experiment, when an experimenter plans to make specific comparisons prior to beginning the collection of data, the hypothesis testing is a priori or planned comparison. He or she plans the comparison in advance because it is of some theoretical interest or practical significance. However, for those comparisons that do not meet these conditions, no prior plans are made to subject these comparisons to specific statistical tests. The resulting test is a posteriori comparison.

The preceding illustration was based on a one-way ANOVA in which there are two or more levels or categories of a single treatment variable. The statistical analysis is designed to test the null hypothesis that the different levels or categories were drawn from the same population. Rejection of the null hypothesis permits us to conclude that the population means from which the samples were drawn are not the same. Therefore, we can conclude that the independent variable exercised a selective effect over the dependent measure.

In Chapter 13, we compared and contrasted two-group independent-samples *t*-tests with two-group correlated-samples *t*-tests. Recall both the advantages and disadvantages of the use of correlated samples (matched pairs or before-after measures on the same subjects). The randomized block one-way ANOVA is an extension of these procedures to designs involving more than two treatment conditions of a single variable and share both the advantages and disadvantages, namely:

Advantages	Disadvantages
Assures investigator that each group is equivalent on the initial ability level of the dependent variable since both members of a pair are roughly equivalent.	There is a loss of degrees of freedom. If the correlation is zero or negligible, the error term is not reduced but a higher value on the test statistic is required for significance.
A source of random error is identified and quantified, thus permitting a corresponding reduction in the error term. A more precise estimate of the treatment effects results.	During the study, for each loss of a member of a matched pair (or a single subject in a before-after design), we lose two observations. High attrition rates can undermine the goals of a study.

The partitioning of the sum of squares in a one-way correlated-measures ANOVA may be diagramed as follows. In point of fact, what would be the within sum of squares in an independent-groups design has been partitioned into two components: blocks SS and residual SS. The residual SS is not usually calculated directly, since blocks SS and the between-group SS can be subtracted from the total SS to yield the residual SS, which is used to calculate the error term.

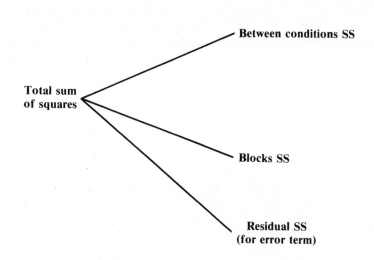

Let's look at a hypothetical study involving three matched groups in a weight-reducing study. Thirty overweight subjects were divided into ten blocks of three SS each. The three lightest SS were randomly assigned to the three conditions and constituted the first block. The same procedures were used to assign the remaining 27 SS to the next nine blocks. Thus, the SS in block 10 were the heaviest at the outset of the study. The following weight losses were recorded on completion of the weight-reducing programs. We'll test the null hypothesis that the weight-reducing programs were drawn from populations with equal means, using $\alpha = .05$, two-tailed test.

Block	X_1	X_1^2	X_2	X_2^2	X_3	X_3^2	Sum of row weight losses
1	7.2	51.84	6.0	36.00	6.4	40.96	19.6
2	9.4	88.36	7.7	59.29	7.9	62.41	25.0
3	8.3	68.89	9.0	81.00	10.3	106.09	27.6
4	11.5	132.25	8.0	64.00	10.0	100.00	29.5
5	15.3	234.09	10.3	106.09	11.3	127.69	36.9
6	14.1	198.81	12.4	153.76	14.0	196.00	40.5
7	17.9	320.41	15.9	252.81	18.9	357.21	52.7
8	23.4	547.56	17.1	292.41	21.2	449.44	61.7
9	21.3	453.69	20.1	404.01	19.4	376.36	60.8
10	26.6	707.56	20.4	416.16	24.8	615.04	71.8
Sum	155.0	2803.46	126.9	1865.53	144.2	2431.20	426.1

Step 1. Find the sum of the weight losses in each treatment condition (column sums) and calculate the means: $\Sigma X_1 = 155.0$, $\Sigma X_2 = 126.9$; $\Sigma X_3 = 144.2$, $\bar{X}_1 = 15.50$, $\bar{X}_2 = 12.69$, $\bar{X}_3 = 14.42$.

Step 2. Square each weight loss and sum to find $\Sigma X_{tot}^2 = 2803.46 + 1865.53 + 2431.20 = 7100.19$.

Step 3. Square the sums of the weight losses and divide by N to find correction term (CT): $CT = (155 + 126.9 + 144.2)^2/30 = (426.1)^2/30 = 6052.04$.

Step 4. Find total sum of squares and df_{tot}: $SS_{tot} = 7100.19 - 6052.04 = 1048.15$; $df_{tot} = 30 - 1 = 29$.

Step 5. Find the sum of squares for treatments, df, and estimated variance: $SS_{bet} = (155^2 + 126.9^2 + 144.2^2)/10 - CT = 60922.25/10 - 6052.04 = 44.19$; $df = 3 - 1 = 2$; $\hat{s}_{bet}^2 = 44.19/2 = 20.09$.

Step 6. Find the block sum of squares, df, and $\hat{s}_{bl}^2$: Square each row sum, add them together, then divide by 3 and subtract CT. $SS_{bl} = (384.16 + 625 + \cdots + 5155.24)/3 - CT = 21,079.09/3 - 6052.04 = 7026.36 - 6052.04 = 974.32$; $df = 10 - 1 = 9$; $\hat{s}_{bl}^2 = 974.32/9 = 108.26$.

Step 7. Find SS_{res} by subtraction, df_{res}, and $\hat{s}_{res}^2$: $SS_{res} = 1048.15 - 974.32$, $df = 30 - 9 - 2 = 18$; $\hat{s}_{res}^2 = 1.87$.

Step 8. Place appropriate values in the Summary Table, calculate relevant F-ratio(s), find the F-ratio(s) required for significance at the chosen level of significance, and test the null hypothesis.

Summary Table

Variable	Sum of squares	df	Estimated variance	F-ratio
Treatment	40.18	2	20.09	10.74
Blocks	974.32	9	108.26	57.89
Residual	33.64	18	1.87	

The obtained F-ratio for treatments is 10.75. The F required for significance at $\alpha = .05$ with df $= 2$ and 9 is 4.26. Thus, we reject the null hypothesis and assert that sample means were not drawn at random from a common population. In other words, there are significant differences among the different weight-reducing programs. To ascertain which conditions differ significantly from which, we apply Tukey's HSD test:

$$\text{HSD} = q_\alpha \sqrt{\hat{s}^2_{\text{res}}/n} = 3.95 \sqrt{1.87/10} = 1.71$$

Thus, any difference between means equal to or greater than 1.71 is significant at the .05 significance level. The table below shows the means and the absolute difference between means in the weight-loss study.

Means and Absolute Differences Between Means

Condition	$\bar{X}_1 = 15.5$	$\bar{X}_2 = 12.69$	$\bar{X}_3 = 14.42$
$\bar{X}_1 = 15.5$		2.81*	1.08
$\bar{X}_2 = 12.69$			1.73*
$\bar{X}_3 = 14.42$			

*Significantly different at the .05 level.

As we can see, both the first and the third conditions produced significantly greater weight loss than the second condition. However, the first and third conditions do not differ significantly from each other.

SELECTED EXERCISES

1. A manufacturer records the number of defective units (out of 100) yielded by three different manufacturing processes. Is there a significant difference among the means? Is one process significantly superior to the others?

Process I		Process II		Process III	
2	4	7	5	5	4
2	5	3	3	1	3
7	4	7		2	2
2	8	9		7	2
5	10	4		3	5
4	9	3		4	4
3		2		4	

2. Three groups of students, randomly assigned to conditions, were taught a subject by three different methods. Do the three groups differ significantly?

Method I		Method II		Method III	
76	79	68	60	70	77
69	60	86	68	87	93
88	73	74	92	92	99
75	81	70	68	60	84
67	62	65	79	95	69

a) Subtract 60 from each score. Conduct an analysis of variance. Compare these results with those obtained earlier.

3. Is there a significant difference among the weights of randomly selected grapefruit from three different fruit growers? (Data are in ounces.)

A	B	C
13.4	12.9	16.0
12.6	13.8	12.1
18.2	12.4	15.0
13.8	13.2	18.2
13.6	12.0	17.0
14.0	13.0	16.3
14.2	12.6	15.4
14.0	14.0	16.0
13.5	13.0	16.0

Employing $\alpha = .01$, test the significance of the difference among the individual means, using Tukey's HSD test.

4. An investigator wished to determine whether there was a significant decrease in the number of errors made as a function of increased trials. Five groups of subjects were tested under five different practice conditions:

Group I = 2 practice trials
Group II = 4 practice trials
Group III = 6 practice trials
Group IV = 8 practice trials
Group V = 10 practice trials

a) Do the groups differ significantly?

I	II	III	IV	V
23	40	34	8	11
39	11	19	21	5
26	28	28	23	12
31	10	9	14	21
44	50	21	15	0
27	26	20	30	16
45	4	10	25	28
40	24	34	13	7
38	28	0	15	15
42	26	15	18	23

b) Employing $\alpha = .05$, determine which of the groups differ significantly, using Tukey's HSD test.

5. Given the following scores on a matching variable, form the subjects into four blocks with three treatments per block.

Subject	Score	Subject	Score
A	120	G	98
B	119	H	105
C	114	I	104
D	115	J	100
E	109	K	117
F	107	L	120

6. Given the following scores on a matching variable, form the subjects into three blocks with four treatments per block.

Subject	Score	Subject	Score
A	22	G	12
B	17	H	18
C	20	I	22
D	17	J	14
E	16	K	21
F	11	L	11

7. Given the following data in which there are three treatment conditions and 21 subjects formed into 7 blocks. Determine the significance of the treatment effects, employing $\alpha = .01$.

Block	Treatment condition X_1	X_2	X_3	Block sum
1	15	13	11	39
2	13	9	10	32
3	12	10	9	31
4	11	13	12	36
5	9	5	7	21
6	8	6	4	18
7	7	5	2	14
	$\sum X_1 = 75$	$\sum X_2 = 61$	$\sum X_3 = 55$	

8. The following scores were obtained in a study employing a randomized block design with three treatment groups and six blocks. Analyze the results using $\alpha = .01$.

Block	Treatment condition X_1	X_2	X_3
1	26	18	23
2	22	14	21
3	18	19	20
4	14	12	16
5	10	6	11
6	8	7	9

Answers:

1. $\sum X_1 = 65$ $\sum X_2 = 43$ $\sum X_3 = 46$

$\sum X_1^2 = 413$ $\sum X_2^2 = 251$ $\sum X_3^2 = 194$

$SS_1 = 88.00$ $SS_2 = 45.56$ $SS_3 = 31.23$

Summary Table

Source of variation	Sum of squares	Degrees of freedom	Variance estimate	F
Between-groups	15.61	2	7.81	1.52
Within-groups	164.79	32	5.15	
Total	180.40	34		

Since the F-ratio is not significant, we shall not test the differences among the means.

2. $SS_1 = 680$ $\qquad$ $SS_2 = 884$ $\qquad$ $SS_3 = 1526.4$

Summary Table

Source of variation	Sum of squares	Degrees of freedom	Variance estimate	F
Between-groups	614.40	2	307.20	2.68
Within-groups	3090.40	27	114.46	
Total	3704.80	29		

3. $SS_A = 20.262$ $\qquad$ $SS_B = 3.209$ $\qquad$ $SS_C = 22.056$

Summary Table

Source of variation	Sum of squares	Degrees of freedom	Variance estimate	F
Between-groups	35.343	2	17.672	9.32
Within-groups	45.527	24	1.897	
Total	80.870	26		

$$HSD = 4.55 \sqrt{\frac{1.897}{3}} = 2.088$$

4. a) $SS_I = 582.5$ $\quad$ $SS_{II} = 1732.1$ $\quad$ $SS_{III} = 1094.0$ $\quad$ $SS_{IV} = 385.6$ $\quad$ $SS_V = 669.6$

Summary Table

Source of variation	Sum of squares	Degrees of freedom	Variance estimate	F
Between-groups	2799.32	4	699.83	7.06
Within-groups	4463.80	45	99.16	
Total	7263.12	49		

b) $$HSD = 4.025 \sqrt{\frac{99.16}{10}} = 12.68$$

5. The blocks would contain the following subjects, not necessarily in the order shown:

Block 1: A, B, L $\qquad$ Block 2: C, D, K

Block 3: E, F, H $\qquad$ Block 4: G, I, J

6. The blocks would contain the following subjects, not necessarily in the order shown:

Block 1: A, C, I, K $\qquad$ Block 2: B, D, E, H $\qquad$ Block 3: F, G, J, L

7. $\bar{X}_1 = 10.71$; $\bar{X}_2 = 8.71$; $\bar{X}_3 = 7.86$

Summary Table

Source of variation	Sum of squares	Degrees of freedom	Variance estimate	F
Between-groups	30.10	2	15.05	8.27
Between blocks	183.81	6	—	
Residual (error)	21.90	12	1.82	
Total	235.81	20		

8. **Step 1.** $\sum X_{tot} = 98 + 76 + 100 = 274$

Step 2. $\dfrac{(\sum X_{tot})^2}{N} = \dfrac{(274)^2}{18} = 4170.89$

Step 3. $\sum X_{tot}^2 = 1844 + 1110 + 1828 = 4782$

Step 4. $SS_{tot} = 4782 - 4170.89 = 611.11$

Step 5. $df_{tot} = 18 - 1 = 17$

Step 6. See Summary Table at Step 16.

Step 7. $SS_{bet} = \dfrac{(98)^2 + (76)^2 + (100)^2}{6} - 4170.89 = 59.11$

Step 8. $df_{bet} = 3 - 1 = 2$

Step 9. See Summary Table at Step 16.

Step 10. $SS_{bl} = \dfrac{(67)^2 + (571)^2 + (571)^2 + (421)^2 + (271)^2 + (24)^2}{3}$
$- 4170.89 = 514.44$

Step 11. $df_{bl} = 6 - 1 = 5$

Step 12. See Summary Table at Step 16.

Step 13. $SS_{res} = 611.11 - (59.11 + 514.44) = 37.56$

Step 14. $df_{res} = 2 \times 5 = 10$

Step 15. See Summary Table at Step 16.

Step 16.

Summary Table

Source of variation	Sum of squares	Degrees of freedom	Variance estimate	F
Between groups	59.11	2	29.56	7.86
Between blocks	514.44	5	—	
Residual (error)	37.56	10	3.76	
Total	611.11	17		

Step 17. $SS_{tot} = SS_{bet} + SS_{bl} + SS_{res}$

$611.11 = 59.11 + 514.44 + 37.56$

$df_{tot} = df_{bet} + df_{bl} + df_{res}$

$= 2 + 5 + 10 = 17$

Step 18. $\hat{s}^2_{bet} = \dfrac{59.11}{2} = 29.56$

Step 19. $\hat{s}^2_{res} = \dfrac{57.56}{10} = 3.76$

Step 20. $F = \dfrac{29.56}{3.76} = 7.86$

Step 21. The critical value of F at $\alpha = .01$ and 2 and 10 degrees of freedom is 7.56. Since obtained F exceeds this value, reject H_0 and assert H_1.

SELF-QUIZ: TRUE-FALSE

Circle T or F.

T   F   1. In a simple analysis of variance, the assumption of homogeneity of variance applies to the total variance.

T   F   2. When $df_{bet} = 1$, $t^2 = F$.

T   F   3. A negative F-ratio indicates a difference in the opposite direction to the one predicted.

T   F   4. The expected value of the F-ratio is 1.00.

T   F   5. In the Student t-ratio, only the denominator gives us some estimate of variability.

T   F   6. $SS_{tot} = SS_W + SS_{bet}$

T   F   7. The within-group sum of squares may sometimes be negative.

T   F   8. If the obtained F-ratio is less than 1.00, we accept H_0.

T   F   9. A significant F-ratio means that the significantly larger within-group variance is due to the operation of the experimental conditions.

T   F   10. Variance estimates are obtained by multiplying the sum of squares by the degrees of freedom.

T   F   11. The within-group variance is analogous to $s_{\bar{x}_1 - \bar{x}_2}$ in the Student t-ratio.

T   F   12. When the parameters are not known, we use the t-ratio to test the significance of the difference between two independent estimates of variance.

T   F   13. The within-group sum of squares equals the sum of the sum of squares obtained within each group.

T   F   14. In a one-way independent-samples ANOVA, the total variance estimate is equal to the sum of the between-group variance estimate and the within-group variance estimate.

T   F   15. The larger the difference between the means is, the larger is the between-group variance.

T   F   16. The F-test is a one-tailed test.

T   F   17. The within-group variance provides an unbiased estimate of the variance of the treatment population from which the sample was randomly selected.

T   F   18. Instead of using Table D with the analysis of variance, we must locate the significant value of F in Table O.

T   F   19. The F-ratio can never be greater than 1.00.

T   F   20. $\dfrac{SS_{tot}}{N - 1} = \dfrac{SS_W}{N - k} + \dfrac{SS_{bet}}{k - 1}$

T   F   21. In a one-way correlated-samples ANOVA design, the within-group sum of squares, divided by its degrees of freedom, provides the error term.

T   F   22. In a randomized block design, subjects are randomly assigned to blocks.

T   F   23. One problem with the randomized block design is that random assignment is not used at any stage of the study.

T   F   24. Given $SS_{tot} = 180$, $df_{tot} = 27$, $SS_{bet} = 40$, $df_{bet} = 6$, $SS_{res} = 80$, $df_{res} = 18$, then $SS_{bl} = 60$.

T   F   25. In True-False Problem 24, there were four blocks.

Answers: (1) F; (2) T; (3) F; (4) T; (5) F; (6) T; (7) F; (8) T; (9) F; (10) F; (11) T; (12) F; (13) T; (14) F; (15) T; (16) F; (17) T; (18) F; (19) F; (20) F; (21) F; (22) F; (23) F ; (24) T; (25) T.

SELF-TEST: MULTIPLE-CHOICE

1. In a simple analysis of variance the assumption of homogeneity of variance applies to:

    a) the variance within the treatment groups
    b) the variance of the means associated with the treatment groups
    c) the total variance
    d) all of the above
    e) none of the above

2. When a variable is "confounded" in an experiment it means that:

    a) it must be varied systematically
    b) it must be randomized with subjects
    c) it cannot be analyzed as a separate source of variance
    d) the treatments are contaminated
    e) all of the above

3. The independent variables in analysis of variance:

    a) must be correlated
    b) may be nominal scales
    c) must be at least interval scales
    d) should be ratio scales
    e) should not be ordinal scales

4. If you obtained an F-ratio equal to .68 with df = 2/20, you would conclude that:

    a) there were no significant differences among the means
    b) you had made an error
    c) the variances were equal
    d) the null hypothesis was rejected
    e) all of the above

5. If the means for each of the treatment groups were identical, the F-ratio would be:

   a) 1.00
   b) zero
   c) a positive number between 0 and 1.00
   d) a negative number
   e) infinite

6. To obtain the between-groups variance estimate, you divide the between-group sum of squares by:

   a) $N - 1$ df
   b) N df
   c) $k - 1$ df
   d) within-groups sum of squares
   e) within-groups variance estimate

7. In the analysis of variance in the two-sample case:

   a) the F-ratio yields the same probability values as the t-ratio
   b) $F = t^2$
   c) it is possible to identify two different bases for estimating the population variance
   d) df = $1/df_W$
   e) all of the above

8. The between-group variance estimate:

   a) is associated with df $= N - k$
   b) reflects the magnitude of the difference among the group means
   c) is analogous to $s_{\bar{x}_1 - \bar{x}_2}$ in the t-ratio
   d) is referred to as the error term
   e) all of the above

9. In multigroup experimental designs, a disadvantage of restricting our statistical analyses to individual comparisons between pairs of experimental conditions is that:

   a) the statistical work is tedious
   b) the risk of a Type II error is increased
   c) the risk of a Type I error is increased
   d) all of the above
   e) none of the above

10. For a given number of degrees of freedom, as the variability among means increases relative to the variability within groups:

   a) the F-ratio is unaffected
   b) the F-ratio increases
   c) the F-ratio decreases
   d) the risk of a Type I error increases
   e) cannot answer without knowing N

11. In multigroup comparisons, if we were alternately to add and subtract a constant from all scores in such a way that the group means remain the same, the F-ratio would:

   a) increase
   b) decrease
   c) remain the same
   d) change in an unpredictable fashion
   e) cannot answer without knowing the value of the constant

12. In multigroup comparisons, if we were to add a constant to all scores, the resulting F-ratio would:

    a) increase
    b) decrease
    c) remain the same
    d) change in an unpredictable fashion
    e) cannot answer without knowing the value of the constant

13. In multigroup comparisons, if we were to subtract a constant from all scores, the resulting F-ratio would:

    a) increase
    b) decrease
    c) remain the same
    d) change in an unpredictable fashion
    e) cannot answer without knowing the value of the constant

14. Which of the following is a true statement?

    a) $SS_{tot} - SS_W = SS_{bet}$ b) $SS_{tot} = SS_W + SS_{bet}$
    c) $SS_{bet} = SS_{tot} - SS_W$ d) all of the above   e) (a) and (b) but not (c)

15. If we were to perform multiple t-tests—without benefit of an analysis of variance—with a several-group experiment, we would be running an increased risk of:

    a) violating the matched-group design
    b) extending the t-distributions beyond their degrees of freedom limitation
    c) a Type I error
    d) not considering the differences attributable to the independent variables
    e) none of the above

16. When comparing between-group and within-group variance estimates, the significance of difference is provided by:

    a) planned comparisons
    b) a posteriori comparisons
    c) the t-ratio
    d) Fisher's F-distributions
    e) none of the above

17. The formula for HSD is given by:

    a) $HSD = q_\alpha \sqrt{\dfrac{\hat{s}_W^2}{N}}$ b) $HSD = \dfrac{\hat{s}_{bet}^2}{\hat{s}_W^2}$ c) $HSD = \dfrac{F}{\sqrt{N-1}}$

    d) $HSD = \sqrt{\dfrac{SS^2}{N_i}}$ e) none of the above

18. In a matched group design, 9 Ss were formed into three blocks on the basis of a preliminary test. The scores were as follows: A, 10; B, 15; C, 36; D, 19; E, 28; F, 5; G, 30; H, 17; I, 8. One block would consist of the following Ss:

    a) C, E, H   b) C, E, G   c) A, B, F   d) C, H, I

19. In a randomized block design, once Ss are assigned to blocks, they are:

    a) assigned from highest to lowest to experimental conditions within blocks
    b) assigned at random within blocks
    c) assigned at random to both experimental conditions and blocks
    d) assigned from lowest to highest to experimental conditions within blocks

e) none of the above

20. Given $SS_{bl_1} = 15$, $SS_{bl_2} = 26$, $SS_{bl_3} = 30$, $k = 5$, then CT equals:

a) 336.07   b) 360.20   c) 1008.20   d) 120.07
e) cannot answer without being given N

Answers: (1) a; (2) c; (3) b; (4) a; (5) b; (6) c; (7) e; (8) b; (9) c; (10) b; (11) b; (12) c; (13) c; (14) d; (15) c; (16) d; (17) a; (18) a; (19) b; (20) a.

Two-Way Analysis of Variance: Factorial Designs, Independent and Correlated Samples

BEHAVIORAL OBJECTIVES

Conceptual Objectives

1. What is the purpose of the two-way ANOVA? When is it applicable?

2. What are the similarities of the one-way and the two-way independent-samples ANOVAs? How do they differ?

3. What are the similarities of the one-way and the two-way correlated-samples ANOVAs? What are the differences?

Procedural Objectives

1. In independent samples designs, know how to partition the total Sum of Squares into Treatments, Interaction, and Within components.

2. In correlated samples designs, know how to partition the Total Sum of Squares into Treatments, Interaction, and Blocks and Residual components.

CHAPTER REVIEW

In Chapter 14, we saw that one-way ANOVA permits us to extend our statistical reach beyond the simple two-group comparison. The development of ANOVA techniques represents an enormous advance for many reasons, one of which is the fact it bring research more in touch with reality. After all, the questions we ask of nature are rarely so simple as "Is there a significant difference between the experimental and the control conditions?" We may wish to evaluate the effects on the dependent variable(s) of many different levels of the quantitative independent variable and attempt to answer such questions as: "If there is an effect of the experimental variable, does it change as a function of the level of that variable? For example, can music be soothing at some sound intensities but aversive at others?" If we're evaluating different categories of a qualitative variable, such as the effects of psychotherapeutic counseling, we may wish to compare any number of different techniques.

But the reach of statistical inquiry is vastly extended by the development of two-way ANOVA and beyond. We may now ask such questions as: "What are the effects of various levels (or categories) of variable A? What are the effects of various levels (or categories) of variable B? Are the effects of A influenced by the level of B and vice versa?" The last question relates to the *interaction* of variables. In some cases, this interaction is of more interest to the investigators than the direct effects of the independent variable. Pharmacists and medical doctors are vitally concerned with the possible interaction of drugs. For example, a given compound may have one effect when taken alone but an entirely different or more pronounced effect when taken in conjunction with another drug. You are probably aware of the joint effects (interaction) of barbiturates and alcohol. Taken alone, each tends to induce sleep. When taken together, the sleep may become permanent.

Thus, many of the questions for which we seek research answers may involve two or more different variables. For example, if we are interested in evaluating the effectiveness of different types of treatment for emotional disturbances, we may use "type of therapy" as one variable and "classification of emotional disorder" as a second variable. Each of these variables may, in turn, be subdivided into two or more categories (psychoanalysis versus person-centered therapy versus rational-emotive therapy, and so forth, and anxiety disorders versus somatoform disorders versus personality disorders, and so on).

Surprisingly, when we go from a one-way ANOVA to a two-way ANOVA (factorial design), the statistical analysis is only slightly changed. The total sum of squares is still partitioned into two components—the treatment combinations sum of squares and the within-group sum of squares. However, the treatment combinations sum of squares is itself partitioned into three components: the A-variable, the B-variable, and the interaction ($A \times B$) of these two variables. The sum of these three components equals the treatment combinations sum of squares. The within-group variance estimate is still used as the error term (denominator) of the three F-ratios that must be computed for each of three different treatment effects, namely, the effects of the A-variable, the B-variable, and the interaction of these two variables. The following table shows one set of data by Manstead et al. (1983) in the study involving emotional contrast. Note that there are two categories of each variable. It is therefore a 2×2 factorial design in which there are four treatment combinations. Note that only the humor manipulation is a true independent variable. The experimenter is not free to assign gender, a naturally occurring variable, to the experimental subjects. The dependent measure is the rating of horror following the viewing of horror scenes. The *lower* the score is the greater is the fright.

	A_1 Horror preceded by humor		A_2 Horror first	
	A_1B_1	$A_1B_1^2$	A_2B_1	$A_2B_1^2$
M	28	784	17	289
	34	1156	31	961
	19	361	33	1089
A	41	1681	33	1089
B_1	14	196	40	1600
L	24	576	24	576
	25	625	41	1681
E	21	441	29	841
	37	1369	19	361
	25	625	30	900
Sum	268	7814	297	9387
	A_1B_2	$A_1B_2^2$	A_2B_2	$A_2B_2^2$
F	11	121	31	961
	29	841	22	484
E	11	121	14	196
	16	256	31	961
M	21	441	27	729
B_2	15	225	32	1024
A	8	64	13	169
	12	144	27	729
L	12	144	16	256
E	17	289	17	289
Sum	152	2646	230	5789

Source: Manstead et al., 1983.

1. Find the sum of the scores for each of the four treatment combinations. In the accompanying example, $A_1B_1 = 268$, $A_2B_1 = 297$, $A_1B_2 = 152$, and $A_2B_2 = 230$. Divide each sum by the number of subjects on which it was based to find the mean for each treatment combination. Thus, $\bar{A}_1\bar{B}_1 = 26.8$, $\bar{A}_2\bar{B}_1 = 29.7$, $\bar{A}_1\bar{B}_2 = 15.2$, and $\bar{A}_2\bar{B}_2 = 23.0$.

At this point it is helpful to prepare a line graph of the means so you can make a preliminary assessment of the effects of each variable and the possible interaction between the two variables.

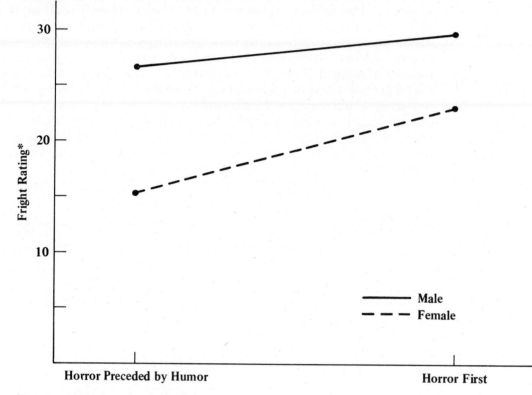

*Low fright rating means high fright.

Inspection of the graph shows that women experienced more fright than men (lower ratings) under both conditions of horror. Also, horror preceded by humor produced greater fright than horror first for both men and women. Since the slopes of the line graphs appear not to differ greatly, we would not expect to find a significant interaction.

1. Add together each group sum to obtain the sum for the total (ΣX_{tot}).

$$\Sigma X_{tot} = 268 + 297 + 152 + 230 = 947$$

2. Square each score and sum. $\Sigma X^2_{tot} = 25{,}636$.

3. Find the total sum of squares: $SS_{tot} = \Sigma X^2_{tot} - (\Sigma X_{tot})^2/N$, $25{,}645 -$ $22{,}420.22 = 3{,}224.78$. Enter into the Summary Table. Note that $(\Sigma X_{tot})^2/N = CT = 22{,}420.22$.

4. Find the total degrees of freedom: $df_{tot} = N - 1 = 39$. Enter into the Summary Table.

5. Find the between-group sum of squares. In the present problem

$$SS_{bet} = \frac{(\Sigma A_1 B_1)^2}{N_{A_1 B_1}} + \frac{(\Sigma A_1 B_2)^2}{N_{A_1 B_2}} + \frac{(\Sigma A_2 B_1)^2}{N_{A_2 B_1}} + \frac{(\Sigma A_2 B_2)^2}{N_{A_2 B_2}} - CT$$

$$= \frac{(268)^2}{10} + \frac{(297)^2}{10} + \frac{(152)^2}{10} + \frac{(230)^2}{10} - 22{,}420.22$$

$$= 25{,}645 - 22{,}420.22 = 3{,}224.78$$

Since N is the same for each treatment combination, there will be less rounding error if the squares of the treatment combinations are summed and then divided by the N in each group (N_K):

$$SS_{bet} = \frac{(\Sigma A_1 B_1)^2 + (\Sigma A_1 B_2)^2 + (\Sigma A_2 B_1)^2 + (\Sigma A_2 B_2)^2}{N_K} - CT$$

$$= \frac{(268)^2 + (297)^2 + (152)^2 + (230)^2}{10} - 22{,}420.22$$

$$= 23{,}603.7 - 22{,}420.22$$

$$= 1183.48$$

Enter into the Summary Table.

6. Find the number of degrees of freedom for the between-group sum of squares.

$$\begin{aligned} df_{bet} &= k - 1 \\ &= 4 - 1 \\ &= 3 \end{aligned}$$

Enter into the Summary Table.

7. Partition the between-group sum of squares into its three components. A = variable, B = variable, and $A \times B$ interaction.

a) $$SS_A = \frac{(\Sigma A_1 B_1 + \Sigma A_1 B_2)^2}{N_{A_1 B_1} + N_{A_1 B_2}} + \frac{(\Sigma A_2 B_1 + \Sigma A_2 B_2)^2}{N_{A_2 B_1} + N_{A_2 B_2}} - CT$$

$$= \frac{(268 + 152)^2}{20} + \frac{(297 + 230)^2}{20} - 22{,}420.22$$

$$= \frac{(420)^2 + (527)^2}{20} - 22{,}420.22$$

$$= 286.33$$

b) $$SS_B = \frac{(\Sigma A_1 B_1 + \Sigma A_2 B_1)^2}{N_{A_1 B_1} + N_{A_2 B_1}} + \frac{(\Sigma A_1 B_2 + \Sigma A_2 B_2)^2}{N_{A_1 B_2} + N_{A_2 B_2}} - CT$$

$$= \frac{(268 + 297)^2}{20} + \frac{(152 + 230)^2}{20} - 22{,}420.20$$

$$= \frac{(565)^2 + (382)^2}{20} - 22{,}420.20$$

$$= 837.23$$

c) $SS_{A \times B} = SS_{bet} - (SS_A + SS_B)$
$$= 1183.48 - (286.23 + 837.23)$$
$$= 60.02$$

Enter the three sums of squares in the Summary Table.

8. Find the number of degrees of freedom for each of the components of the between-group estimates.

   a)
   $$df_A = A - 1$$
   $$= 2 - 1$$
   $$= 1$$

   b)
   $$df_B = B - 1$$
   $$= 2 - 1$$
   $$= 1$$

   c)
   $$df_{A \times B} = df_A \times df_B$$
   $$= 1 \times 1$$
   $$= 1$$

   Enter in the Summary Table the degrees of freedom for each component of the between-group estimate.

9. Calculate the within-group (error) sum of squares (by subtraction) and df_W.

   $$SS_W = SS_{tot} - SS_{bet}$$
   $$= 3224.78 - 1183.48$$
   $$= 2041.3$$
   $$df_W = N - k$$
   $$= 40 - 4 = 36$$

   Enter the within-group sum of squares into the Summary Table.

10. Find variance estimates for the treatment variables by dividing their sums of squares by their respective degrees of freedom.

    a) $\hat{s}_A^2 = \dfrac{SS_A}{df_A} = \dfrac{286.23}{1} = 286.23$

    b) $\hat{s}_B^2 = \dfrac{SS_B}{df_B} = \dfrac{837.23}{1} = 837.23$

    c) $\hat{s}_{A \times B} = \dfrac{SS_{A \times B}}{df_{A \times B}} = \dfrac{60.02}{1} = 60.02$

    Enter all three into the Summary Table.

11. Find the estimated within-group variance estimate (error).

    $$\hat{s}_W^2 = \frac{SS_W}{df_W} = \frac{2041.3}{36} = 56.7$$

    Enter into the Summary Table.

12. Calculate the F-ratio for each of the treatment variables and their interaction.

    a) $F_A = \hat{s}_A^2 / \hat{s}_W^2 = 286.23/56.7 = 5.048$

    b) $F_B = \hat{s}_B^2 / \hat{s}_W^2 = 837.23/56.7 = 14.766$

    c) $F_{A \times B} = \hat{s}_{A \times B}^2 / \hat{s}_W^2 = 60.02/56.7 = 1.059$

13. Using $\alpha = .05$, refer to Table D in the appendix for critical values at the .05 level. Note that since all treatment effects involve df = 1, a single critical value suffices to evaluate all three effects. This would not be the case if there

were more than two levels or categories of either or both treatment variables. We find that at $\alpha = .05$, $F \geq 4.11$ is required for significance at 1 and 36 degrees of freedom.

Since the F-ratios for both the treatment effects, A and B, exceed the critical value, we may reject H_0 in both instances. However, we fail to reject H_0 associated with the interaction of these two variables.

We may conclude that (1) humor preceding horror scenes produces significantly greater fright, regardless of gender, than horror scenes shown first and (2) compared to men, women rated all horror scenes as more frightful, whether or not preceded by scenes of humor.

Summary Table for a 2 × 2 ANOVA

Variable	Sum of squares		df	Variance estimate	F-ratio
Treatment combination	1183.48		3		
Variable A		286.23	1	286.23	5.048
Variable B		837.23	1	837.23	14.766
Interaction		60.02	1	60.02	1.059
Within (error)	2041.31		36	56.703	
Total	3224.79		39		

You may have observed the parallel structure of Chapters 13, 14, and 15 in the text. In each chapter, independent-samples designs are followed by correlated-samples designs, as shown in the following list:

Chapter 13	t-test for independent samples	t-test for correlated samples
Chapter 14	one-way ANOVA for independent samples	one-way ANOVA for correlated samples
Chapter 15	two-way ANOVA for independent samples	two-way ANOVA for correlated samples

Two-way ANOVA for correlated samples presents the same advantages and disadvantages as other correlated-samples designs. However, since a given block may include a large number of matched Ss in a two-way design, it may be difficult to achieve effective matching across so many treatment combinations. Moreover, the effect of subject attrition may be quite pronounced and should be carefully evaluated prior to opting for this design.

Let's look at some data derived from a hypothetical study involving a 2×3 factorial randomized block design:

	A_1			A_2			Block sum
	B_1	B_2	B_3	B_1	B_2	B_3	
Block 1	2	4	4	4	2	1	17
Block 2	4	6	9	5	3	3	30
Block 3	5	8	11	8	5	2	39
Block 4	7	9	13	7	4	3	43
Block 5	8	9	12	10	6	4	49
Sum	26	36	49	34	20	13	178

Step 1. Find the column sums and calculate the mean of each column: $\overline{A_1B_1} = 5.2$; $\overline{A_1B_2} = 7.2$; $\overline{A_1B_3} = 9.8$; $\overline{A_2B_1} = 6.8$; $\overline{A_2B_2} = 4.0$; $\overline{A_2B_3} = 2.6 = 6.20$. It is frequently informative to chart these means at this point:

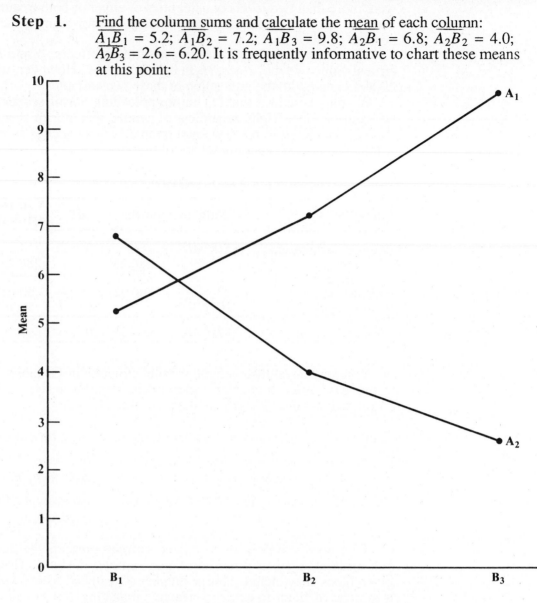

By inspection, we see that the A_1 means are higher than the A_2 means at levels B_2 and B_3 but lower at B_1. There appears to be little difference among the B means at any level of A. But most striking is the crossing of the two lines, suggesting a substantial interaction between the A and B variables. We now continue with the analysis to ascertain if any of these differences achieve statistical significance.

Step 2. Find the total sum of squares and total degrees of freedom: $SS_{tot} = 2^2 + 4^2 + \cdots + 4^2 - CT$, in which $CT = 178^2/35 = 31{,}684/30 = 1056.13$. $SS_{tot} = 1350 - 1056.13 = 293.87$; $df_{tot} = N - 1 = 29$.

Step 3. Calculate the blocks sum of squares and degrees of freedom: $SS_{bl} = (17^2 + 30^2 + \cdots + 49^2)/4 - CT$. $SS_{bl} = 1160 - 1056.13 = 103.87$; $df_{bl} = bl - 1 = 4$.

Step 4. Calculate the treatment sum of squares and degrees of freedom: $SS_{treat} = (26^2 + 36^2 + \cdots + 20^2)/5 - CT$. $SS_{treat} = 1219.60 - 1056.13 = 163.47$; $df_{tr} = treat - 1 = 5$.

Step 5. Find the residual sum of squares and degrees of freedom by subtraction: $SS_{res} = SS_{tot} - SS_{bl} - SS_{treat} = 293.87 - 103.87 - 163.47 = 26.53$; $df_{res} = df_{tot} - df_{bl} - df_{tr} = 29 - 4 - 5 = 20$.

Step 6. Prepare to partition the treatment sum of squares into: SS_A, SS_B, and $SS_{A \times B}$ by setting up the following table of cell and marginal sums:

	Variable		
	A_1	A_2	B sums
B_1	26	34	60 $n = 10$
B_2	36	20	56 $n = 10$
B_3	49	13	62 $n = 10$
A sums	111 $n = 15$	67 $n = 15$	178

Step 7. Find the sum of squares and degrees of freedom for the A variable: $SS_A = (111^2 + 64^2)/15 - CT = 64.53$; $df_A = 2 - 1 = 1$.

Step 8. Find the sum of squares and degrees of freedom for the B variable: $SS_B = (60^2 + 56^2 + 62^2)/10 - CT = 1.87$; $df_B = 3 - 1 = 2$.

Step 9. Obtain the interaction sum of squares and degrees of freedom by subtraction: $SS_{A \times B} = SS_{treat} - SS_A - SS_B = 163.47 - 64.53 - 1.87 = 97.07$; $df_{A \times B} = df_{treat} - df_A - df_B = 5 - 1 - 2 = 2$.

Step 10. Prepare the Summary Table and enter the relevant values:

Summary Table

Variable	Sum of squares	df		Variance estimate	F-ratio
Treatment combination	163.47	5			
Variable A		64.53	1	64.53	48.52
Variable B		1.87	2	.93	.17
Interaction		97.07	2	48.53	8.93
Blocks	103.87	4			
Residual (error)	26.53	20		1.33	
Total	293.87	29			

Step 11. Obtain the F-ratio for the A-variable and test significance at the .05 level at 1 and 20 df. $F = 64.53/1.33 = 48.52$. Referring to Table D we find that an F-ratio of 4.35 or greater is required for significance. Decision: Reject H_0. There is a significant effect of the A-variable.

Step 12. Obtain the F-ratio for the B-variable and test significance at the .05 level at 2 and 20 df. $F = 3.49$ or greater is required for significance. Decision: Fail to reject H_0. There is not a significant effect of the B-variable.

Step 13. Obtain the F-ratio for the $A \times B$ interaction and test the significance at the .05 level at 2 and 20 df. $F = 3.49$ or greater is required for significance. Decision: Reject H_0. There is a significant interaction between variables A and B. This interaction means that the level of variable A is dependent on the level of variable B. Thus, the conclusion in **11**, above, must be modified. We cannot conclude that the effect of variable A is the same over all levels of variable B. In other words, there is not a *main* effect of the A-variable but rather a *simple* effect that depends on the level of B.

SELECTED EXERCISES

1. How many treatment combinations are there in the following independent-samples factorial designs?

    a) 4×5 b) 2×4 c) 3×4 d) 5×6

2. How many treatment combinations are there in the following randomized block factorial designs?

    a) 4×5 b) 2×4 c) 3×4 d) 5×6

3. Show all the treatment combinations in a 4×5 independent-samples factorial design.

4. Show the assignment of degrees of freedom in a 3×4 independent-samples factorial design in which there are two observations per cell or treatment combination.

5. Show the assignment of degrees of freedom in a 3×3 randomized block factorial design in which there are 4 blocks of 9 SS each.

6. Following are the data in a 3×3 independent-samples factorial design. Conduct an ANOVA and determine if there is a significant effect of the A-variable, the B-variable, or an interaction of the two variables. Use $\alpha = .01$. If warranted, apply Tukey's HSD test of the significance of the difference between means, using $\alpha = .05$.

	A_1			A_2			A_3		
	B_1	B_2	B_3	B_1	B_2	B_3	B_1	B_2	B_3
	2	3	6	1	2	5	1	5	6
	4	5	7	3	5	8	5	8	10
	5	8	9	6	7	6	3	7	12
	7	11	12	7	9	11	7	9	11
Treatment combination	A_1B_1	A_1B_2	A_1B_3	A_2B_1	A_2B_2	A_2B_3	A_3B_1	A_3B_2	A_3B_3
Σ	18	27	34	17	23	30	19	29	39

$$\Sigma X_T = 236$$

7. Following are the data from a 2×3 randomized block factorial design in which there are four blocks of six subjects each. Conduct the appropriate ANOVA, using $\alpha = .05$. If warranted, apply Tukey's HSD test and ascertain if there are significant differences between pairs of means, using $\alpha = .05$.

| | A_1 | | | A_2 | | |
	B_1	B_2	B_3	B_1	B_2	B_3
Block 1	1	5	1	8	6	11
Block 2	6	8	6	7	9	9
Block 3	7	3	2	9	12	13
Block 4	9	7	5	11	8	7

Answers:

1. a) 20   b) 8   c) 12   d) 30

2. Same answers as in Selected Exercise 1.

3. a) $A_1B_1, A_1B_2, A_1B_3, A_1B_4, A_1B_5$

   b) $A_2B_1, A_2B_2, A_2B_3, A_2B_4, A_2B_5$

   c) $A_3B_1, A_3B_2, A_3B_3, A_3B_4, A_3B_5$

   d) $A_4B_1, A_4B_2, A_4B_3, A_4B_4, A_4B_5$

4. $df_{tot} = 23$, $df_{tr} = 11$ ($df_A = 2$, $df_B = 3$, $df_{A \times B} = 6$); $df_W = 12$

5. $df_{tot} = 35$; $df_{bl} = 3$; $df_{tr} = 8$ ($df_A = 2$, $df_B = 2$, $df_{A \times B} = 4$); $df_{res} = 24$.

6. $SS_{tot} = 1842 - (236)^2/36 = 1842 - 1547.11 = 294.89$; $df_{tot} = 36 - 1 = 35$; $SS_{treat} = 1662.50 - 1547.11 = 115.39$; $df_{treat} = 9 - 1 = 8$; $SS_A = 1559.17 - 1547.11 = 12.06$; $df_A = A - 1 = 2$; $SS_B = 1647.17 - 1547.11 = 100.06$; $df_B = B - 1 = 2$; $SS_{A \times B} = 115.39 - 12.06 - 100.06 = 3.27$; $df_{A \times B} = (A - 1)(B - 1) = 2 \times 2 = 4$; $SS_W = 294.89 - 115.39 = 179.5$; $df_W = N - treat = 36 - 9 = 27$.

Source of Variation	Sum of squares	df	Variance estimate	F-ratio
Treatment combinations	115.39	8		
Variable A	12.06	2	6.03	.91
Variable B	100.06	2	50.03	7.52
$A \times B$	3.27	4	0.82	0.12
Within-group (error)	179.50	27	6.65	
Total	294.89	35		

Since the interaction and A variable are not statistically significant but the B variable is, it is appropriate to apply the HSD test to the B-variable:

	$\bar{X}_{B_1} = 4.50$	$\bar{X}_{B_2} = 6.58$	$\bar{X}_{B_3} = 8.58$
$\bar{X}_{B_1} = 4.50$	—	2.08	4.08
$\bar{X}_{B_2} = 6.58$		—	2.00
$\bar{X}_{B_3} = 8.58$			—

The n per condition is 12. Thus

$$HSD = 4.50 \sqrt{\frac{6.65}{12}}$$

$$= (4.50)(0.74)$$

$$= 3.33$$

Only the pairwise comparison between the means of B_1 and B_3 is statistically significant.

7. $SS_{tot} = 179.333$; $df_{tot} = 24 - 1 = 23$; $SS_{treat} = 95.333$; $df_{treat} = 6 - 1 = 5$; $SS_A = 80.667$; $df_A = A - 1 = 1$; $SS_B = 1.084$; $df_B = B - 1 = 2$; $SS_{A \times B} = 13.583$; $df_{A \times B} = (A - 1)(B - 1) = 2 \times 2 = 4$; $SS_{bl} = 8.33$; $df_{bl} = 4 - 1 = 3$; $SS_{res} = 75.667$; A means, $\bar{A}_1 = 5.500$; $\bar{A}_2 = 9.167$; B means, $\bar{B}_1 = 7.625$, $\bar{B}_2 = 7.250$, $\bar{B}_3 = 7.125$.

Summary Table

Variable	Sum of squares	df	Variance estimate	F-ratio
Treatment combination	95.333	5		
Variable A	80.667	1	80.667	15.99
Variable B	1.084	2	.542	.11
Interaction	97.07	2	6.791	1.35
Blocks	8.333	3		
Residual (error)	75.667	15	5.044	
Total	179.333	23		

Only the A-variable is statistically significant. Since there are only two A means, an HSD test is not necessary. The difference between these two means is statistically significant, as shown by the significant F-ratio. If only the B-variable had been statistically significant, Tukey's HSD test of significance among B means would have been appropriate. Had only the interaction been significant, the following HSD comparisons for simple effects would have been appropriate: $\overline{A_1B_1}$ versus $\overline{A_1B_2}$ and $\overline{A_1B_3}$, $\overline{A_2B_1}$ versus $\overline{A_2B_2}$ and $\overline{A_2B_3}$.

SELF-TEST: MULTIPLE CHOICE

1. A 5×6 factorial design involves:

   a) 11 different treatment levels
   b) five independent variables at six treatment levels
   c) five levels of one treatment variable and six levels of another
   d) five dependent and six independent variables

2. In a 7×8 factorial design, the number of treatment combinations is:

   a) 28   b) 56   c) 42   d) 15

3. If a subject received the fourth level of treatment A and the second level of treatment B in a 5×6 factorial design, the treatment combination would be:

   a) A_4B_2 b) A_5B_6 c) 4×2 d) 5×6

4. In a two-way analysis of variance, the treatment combinations sum of squares may be partitioned into the following three components:

   a) total, treatment combinations, within-group
   b) total, within-group, A-variable
   c) A-variable, B-variable, within-group
   d) A-variable, B-variable, $A \times B$ interaction

5. In a two-way analysis of variance, each observation yields information concerning how many treatment effects and their interactions?

   a) one   b) two   c) three   d) four

6. In a 3×3 factorial design, the following scores were obtained for each treatment combination: 8, 12, 9, 14, 16, 12, 10, 5, 3. The n in each treatment combination equaled two. SS_{treat} equals:

   a) 590.50   b) 113.00   c) 48.22   d) 2325.5

7. In Multiple-Choice Problem 6, the sum of each treatment level of the B-variable was 35, 30, and 16. The sum of squares for B was:

   a) 99.94   b) 826.0   c) 32.33   d) 28.74

8. In a 6×7 factorial design in which $N = 210$, df_W equals:

   a) 168   b) 42   c) 209   d) 30

9. In a 4×5 factorial design in which $N = 100$, df_A equals:

   a) 4   b) 99   c) 80   d) 3

10. In a 4×4 factorial design in which $N = 64$, $df_{A \times B}$ equals:

    a) 9   b) 16   c) 63   d) 3

11. In a 3×4 factorial design in which $N = 72$, df_r equals:

    a) 9   b) 71   c) 60   d) 12

12. Given $SS_{treat} = 205$, $SS_A = 35$, $SS_B = 90$, $SS_{tot} = 450$. $SS_{A \times B}$ equals:
    a) 245   b) 80   c) 325   d) 655

13. Given $SS_{tot} = 60$, $SS_{treat} = 15$, $SS_A = 5$, $SS_{A \times B} = 3$. SS_B and SS_W equal, respectively:

    a) 7, 45   b) 8, 45   c) 7, 75   d) 23, 75

14. Given $SS_{treat} = 180$, $SS_A = 70$, $SS_B = 60$, $SS_{A \times B} = 50$, $SS_W = 140$, and $df_A = 2$, $df_B = 3$, $df_{A \times B} = 9$, and $df_W = 60$. $\hat{s}_A^2$ equals:

    a) 0.50   b) 15.02   c) 35   d) 23.33

15. Given the same data as in Multiple-Choice Problem 14, $\hat{s}_{A \times B}^2$ equals:

    a) 16.67   b) 25   c) 5.56   d) 2.39

16. Following are scores made by subjects in a 3×3 factorial experiment. Conduct the appropriate analysis of variance and show conclusions warranted by the analysis. Use $\alpha = .05$.

A_1			A_2			A_3		
B_1	B_2	B_3	B_1	B_2	B_3	B_1	B_2	B_3
8	10	12	5	9	16	6	10	17
1	7	9	3	10	11	5	7	9
7	9	11	8	10	12	4	11	12
9	7	5	8	6	4	10	8	6
4	8	15	6	8	10	8	10	12
12	14	15	10	12	13	13	14	14

Answers: (1) c;   (2) b;   (3) a;   (4) d;   (5) c;   (6) b;   (7) c;   (8) a;   (9) d;   (10) a;
(11) c;   (12) b;   (13) a;   (14) c;   (15) c.
(16):

Step 1. $SS_{tot} = 5274 - 4629.63 = 644.37$

Step 2. $df_{tot} = 54 - 1 = 53$

Step 3.   See Summary Table at Step 19.

Step 4. $SS_{treat} = \dfrac{(41)^2 + (55)^2 + \cdots + (70)^2}{6} - 4629.63$
$$= 1798.67 - 4629.63 = 169.04$$

Step 5. $df_{treat} = 9 - 1 = 8$

Step 6.   See Summary Table at Step 19.

Step 7. $SS_A = \dfrac{(163)^2 + (161)^2 + (176)^2}{18} - 4629.63$
$$= 4637 - 4629.63 = 7.37$$

Step 8. $df_A = 3 - 1 = 2$

Step 9.   See Summary Table at Step 19.

Step 10. $SS_B = \dfrac{(127)^2 + (170)^2 + (203)^2}{18} - 4629.63$
$$= 4791 - 4629.63 = 161.37$$

Step 11. $df_B = 3 - 1 = 2$

Step 12.   See Summary Table at Step 19.

Step 13. $SS_{A \times B} = 169.04 - (7.37 + 161.37) = .30$

Step 14. $df_{A \times B} = 2 \times 2 = 4$

Step 15.   See Summary Table at Step 19.

Step 16. $SS_W = 644.37 - 169.04 = 475.33$

Step 17. $df_W = 54 - 9 = 45$

Step 18.   See Summary Table at Step 19.

Step 19.

Summary Table

Variable	Sum of squares	df	Variance estimate	F-ratio
Treatment combinations		169.04	8		
Variable A	7.37	2	3.68	.35
Variable B	161.37	2	80.68	7.64
A × B	.30	4	.08	0.01
Within (error)	475.33	45	10.56	
Total	644.37	53		

Step 20. $SS_{treat} + SS_W = SS_{tot}$
$$169.04 + 175.33 = 644.37$$
$$SS_A + SS_B + SS_{A \times B} = SS_{treat}$$
$$7.37 + 161.37 + .30 = 169.04$$

Step 21. $\hat{s}^2_{A \times B} = \dfrac{.30}{4} = .08$

Step 22. $\hat{s}^2_B = \dfrac{161.37}{2} = 80.68$

Step 23. $\hat{s}^2_A = \dfrac{7.37}{2} = 3.68$

Step 24. $\hat{s}^2_W = \dfrac{475.33}{45} = 10.56$

Steps 25, 26. $F_{A \times B} = \dfrac{.08}{10.56} = .01$, df 4/45, not significant

Steps 27, 28. $F_B = \dfrac{80.68}{10.56} = 7.64$, df 2/45, significant

Steps 29, 30. $F_A = \dfrac{7.08}{10.56} = .35$, df 2/45, not significant

Since the B-variable yields statistical significance, we apply the Tukey HSD test:

$\bar{X}_{B_1} = \dfrac{127}{18} = 7.06$, $\bar{X}_{B_2} = \dfrac{170}{18} = 9.44$, $\bar{X}_{B_3} = \dfrac{203}{18} = 11.28$.

Step 31.

	$\bar{X}_{B_1} = 7.06$	$\bar{X}_{B_2} = 9.44$	$\bar{X}_{B_3} = 11.28$
$\bar{X}_{B_1} = 7.06$		2.38	4.22
$\bar{X}_{B_2} = 9.44$			1.84
$\bar{X}_{B_3} = 11.28$			

Step 32. $q_\alpha = .05$ 3.43, by linear interpolation.

Step 33. $\text{HSD} = 3.43\sqrt{\dfrac{10.56}{18}} = 2.64$

Step 34.   Only the difference between conditions B_1 and B_3 is significant at the .05 level.

PART IV

INFERENTIAL STATISTICS: NONPARAMETRIC TESTS OF SIGNIFICANCE

16

Power and Power Efficiency of a Statistical Test

BEHAVIORAL OBJECTIVES

Conceptual Objectives

1. So far as the concept of power is concerned, what is an ideal statistical test? In words and in symbolic notation, give the definition of power.

2. List the four factors that have a direct effect on the power of a statistical test. Precisely state the relationship of each factor to the power of a test. Given specific situations of each factor and changes in these conditions, relate the effect on power. Familiarize yourself with the characteristics of each factor.

3. Considering the mutually exclusive situations offered by the null hypothesis, what are the probabilities associated with Type I and Type II errors? When is the concept of power applicable?

4. Define power efficiency. How does it relate to parametric and nonparametric tests?

Procedural Objectives

1. In both the one-sample and the two-sample cases, given the appropriate data, calculate the power of a test when σ and μ are known. *Note:* This procedure would involve the use of z-scores and the normal curve.

2. When the parameters are known, calculate the power of a test and probability of a Type II error with correlated samples.

3. Compute the power efficiency of a nonparametric test in relationship to a competing parametric test.

CHAPTER REVIEW

Although we have mentioned the power of a statistical test before, now is the time for an in-depth discussion of the concept. The power of a test can be defined very simply as the *probability of rejecting the null hypothesis when it is actually false*. From this definition of power, we can immediately seize upon two qualities of the power of a test.

First, power is expressed as a probability. Second, the null hypothesis must be false before the concept of power is even applicable. The implications of these two points are that if we know the null hypothesis to be true in a given test, we need not discuss the power of that test. It simply is not applicable. Also, since power is a probability value, it can never exceed 1.00, nor will it be a value less than 0.00. However, as we know, we can convert probability values to percentages. In this case power would be expressed as a percentage between 0% and 100%.

Since the definition of power is the probability of rejecting the null hypothesis when it is false, power can be symbolically stated as

$$\text{Power} = 1 - \text{probability of a Type II error}$$

However, the probability of a Type II error is given by the quantity β. Therefore

$$\text{Power} = 1 - \beta$$

Suppose in a test of significance we accept H_0 when it is false. If β in this instance has a calculated value of .76, the power of the test is $1 - .76 = .24$. As you see, when the probability of making a Type II error is high, the power of the test is correspondingly low.

If we know the population parameters in a problem, we can use z-scores and the standard normal distribution to determine probability values. We can calculate the probability of making a Type II error and the power of the test. We can obtain all of these probability values through the use of the standard normal distribution.

Let us suppose we know that the mean running speed of a group of subjects under specified conditions is 40, $\sigma = 10$. We wish to test whether a specific change in these conditions will change the mean running speed ($H_0:\mu = \mu_0 = 40$). Suppose that we use a group of 25 subjects. Using $\alpha = .05$, two-tailed test, what is the power of the test if the true mean equals 42 ($\mu = \mu_1 = 42$)?

1. Calculate the value of $\sigma_{\bar{X}}$:

$$\sigma_{\bar{X}} = \frac{\sigma}{\sqrt{N}} = \frac{10}{5} = 2.0$$

2. Determine the critical value of $\bar{X}$ at $\alpha = .05$:

$$\begin{aligned} \bar{X} &= (z_{.01})(\sigma_{\bar{X}}) + \mu_0 \\ &= (\pm 1.96)(2) + 40 \end{aligned}$$

$$\begin{aligned} \text{Upper value } \bar{X} &= (+1.96)(2) + 40 = 43.92 \\ \text{Lower value } \bar{X} &= (-1.96)(2) + 40 = 36.08 \end{aligned}$$

3. Determine the probability of obtaining these critical values in the true sampling distribution under H_1. The upper critical value has a z-score of

$$\begin{aligned} z &= \frac{43.92 - 42}{2.0} \\ &= .96 \end{aligned}$$

Referring to column C (Table A in the text), we find that $p = .1685$. The lower critical value has a z-score of

$$\begin{aligned} z &= \frac{36.08 - 42}{2.0} \\ &= -2.96 \end{aligned}$$

The proportion of area beyond $z = -2.96$ is 0.0015.

The power of the test equals $.1685 + .0015$ or .17. Expressed as a percentage, the probability of correctly rejecting a false null hypothesis is only 17%. The probability, then, of mistakenly accepting a false H_0 is 83%.

The calculation of power in the two-sample case involves essentially the same procedures as in the one-sample case.

Let us suppose that we have two populations with the following parameters:

$$\begin{array}{ll} \mu_1 = 68 & \mu_2 = 65 \\ \sigma_1 = 4 & \sigma_2 = 4 \end{array}$$

Suppose that we draw a sample of 16 cases from each population. What is the power of the test ($H_0:\mu_1 \le \mu_2 = 0$), using $\alpha = .01$, one-tailed test ($z = 2.33$)?

1. Calculate the value of $\sigma_{\bar{X}_1 - \bar{X}_2}$:

$$\sigma_{\bar{X}_1} = \frac{\sigma_1}{\sqrt{N_1}} = \frac{4}{\sqrt{16}} = 1.0$$

$$\sigma_{\bar{X}_2} = \frac{\sigma_2}{\sqrt{N_2}} = \frac{4}{\sqrt{16}} = 1.0$$

$$\sigma_{\bar{X}_1 - \bar{X}_2} = \sqrt{\sigma_{\bar{X}_1}^2 + \sigma_{\bar{X}_2}^2} = \sqrt{2} = 1.41$$

2. Determine the critical value of $\bar{X}_1 - \bar{X}_2$:

$$\bar{X}_1 - \bar{X}_2 = (2.33)(1.41) = 3.29$$

3. Determine the probability of obtaining this critical value in the true sampling distribution under H_1:

$$= \frac{3.29 - 3.00}{1.41}$$
$$= .21$$

Referring to column C (Table A in the text), we find that the area beyond $z = .21$ is .4168. Thus, the power of the test equals 41.68%.

Many factors have an influence on the power of a test. Four of these that we shall discuss are sample size, α-level, the nature of the alternative hypothesis, and the nature of the test: parametric versus nonparametric.

In general, we can say that the larger the sample size is, the higher is the power of the test. If we select two samples ($N_1 = 10$, $N_2 = 60$) from a single population, we would expect the power of the test on the first sample to be less than on the second sample of data.

A simple statement of the effect of α-level on power is that the higher the α-level that is chosen, the greater is the power of the test. In this instance, "higher" implies a larger value of α. For example, an α-level of .05 would yield more power than a .01 α-level on the same test.

Another factor affecting the amount of power a test has is the nature of the alternative hypothesis. More often than not, a one-tailed test is more powerful than the corresponding two-tailed test. This is to say that a directional alternative hypothesis produces a more powerful test than a nondirectional test would under the same circumstances. There is however, an exception to this rule: If the parameter actually lies in the opposite direction to the one predicted, the two-tailed test is more powerful. For clarity, consider the case in which the statement of the alternative hypothesis is $\mu_1 > 39$. If, in fact, the population mean is 32, a nondirectional H_1 would have given us a more powerful test than the hypothesis $\mu > 39$. If we are accurate in the direction of our predictions, the one-tailed test is more powerful than the two-tailed test.

Our fourth factor is the nature of the test itself, that is, whether it is parametric or nonparametric. For any given N, when the underlying assumptions of the parametric test are met, the power of the parametric test is greater than its nonparametric counterpart. If you were told that two tests—one parametric, the other nonparametric—were conducted on the same sample of data drawn from a normally distributed population, you should expect the nonparametric tests to have less power than the parametric one.

Although it is true that the power of a nonparametric test is not so great as the power of a parametric test if N is held constant, by enlarging the sample size under the nonparametric condition, we can make the nonparametric test just as powerful as the parametric one. Thus, by adding to the sample size, the nonparametric test can become equal in power to its parametric equivalent. When we increase N to make one test as powerful as a competing test, we are applying the concept of power efficiency. By definition, power efficiency is the increase in sample size that is necessary to make one test as powerful as a competing test.

Consider the situation in which we have two tests, A and B. Since A is a parametric test and B is not, when $N = 10$ for both samples, A is a more powerful test than B. However, if we increase the sample size for test B to 15, we have equal power for the two tests. How do we determine exactly what the power efficiency of test B is? The following formula should answer this question:

$$\text{Power efficiency of test B} = 100 \frac{N_a}{N_b}\%$$

If we substitute our sample sizes into the formula, we find

$$\text{Power efficiency of test B} = 100 \left(\frac{10}{15}\right)\%$$

or 67%. For every two cases of test A, we will need three cases of test B to achieve power equivalency between tests.

You may wonder: "Why increase the N for a less powerful nonparametric test when an identical increase in the parametric counterpart will achieve even greater power?" The answer to this question often involves both economic and time considerations. If there are unlimited time and funds, most would agree that the more powerful test, if available, should be used. However, this test may be expensive and time-consuming to develop and/or to administer. If an answer is needed within a short time or if there are insufficient funds to develop or administer a ratio- or interval-scaled dependent variable, we may be forced to consider a less precise but more available ordinal- or nominal-dependent measure. Then the question becomes relevant: "How much of an increase in N will make the nonparametric counterpart as powerful as the parametric alternative?" If the cost and time analysis reveal that the research can be completed in a timely fashion and within budget, the research may have little choice but to opt for the nonparametric alternative.

SELECTED EXERCISES

1. An investigator was interested in the effect of a new drug on the reaction times of nine Ss. The mean reaction time of the Ss was found to be 20.8 seconds. If the true mean and standard deviation are 15.5 and 9 seconds, respectively, what is the power of the test:

   a) employing $\alpha = .01$, two-tailed test?
   b) employing $\alpha = .05$, two-tailed test?

2. Suppose that the investigator in Selected Exercise 1 stated the following alternative hypothesis: $H_1 : \mu < 20.8$. Employing $\alpha = .05$, determine the power of this one-tailed test.

3. Employing $\alpha = .01$ for the data in Problem 1, what is the power of the test if the following alternative hypothesis was stated: $H_1 : \mu < 20.8$? Compare your result with that of Selected Exercise 2.

4. Given two normal populations:
   $\mu_1 = 68$ $\mu_2 = 62$
   $\sigma_1 = 10$ $\sigma_2 = 10$

Employing $\alpha = .01$, one-tailed test $(H_1:\mu_1 > \mu_2)$, determine the power of the test when

a) $N_1 = N_2 = 4$ b) $N_1 = N_2 = 16$ c) $N_1 = N_2 = 64$ d) $N_1 = N_2 = 100$

5. Given two normal populations:
$\mu_1 = 1.5$ $\mu_2 = 1.3$
$\sigma_1 = .6$ $\sigma_2 = .6$
Employ $\alpha = .05$, two-tailed test. If a sample of 36 cases is drawn from each population, find

a) the probability of a Type I error
b) the probability of a Type II error
c) the power of the test

6. The power efficiency of test A relative to test B is 75%. If we employed 36 subjects with test B, what is the N required to achieve equal power with test A?

7. In order to take test A as powerful as test B, we must employ 5 subjects with test A for every 3 subjects with test B. What is the power efficiency of test A relative to test B?

8. When is a statistical test for independent groups more powerful than a test for correlated groups?

Answers:

1. a) $\sigma_{\bar{X}} = 9/\sqrt{9} = 3$; upper value of $\bar{X} = 20.8 + (2.58)(3) = 25.54$; lower value of $\bar{X} = 20.8 - (2.58)(3) = 13.06$; $z_1 = (13.06 - 15.5)/3 = -.81$; $z_u = (28.54 - 15.5)/3 = 4.35$; area beyond $z_1 = .2090$; area beyond z_u is negligible. Thus, power = 20.90%.

   b) $\sigma_{\bar{X}} = 9/\sqrt{9} = 3$; upper value of $\bar{X} = 20.8 + (1.96)(3) = 26.68$; lower value of $\bar{X} = 20.8 - (1.96)(3) = 14.92$; $z_1 = (14.92 - 15.5)/3 = -0.19$; $z_u = (26.68 - 15.5)/3 = 3.73$; area beyond $z_1 = .4247$; area beyond z_u is .0001. Thus, power = 42.48%.

2. Lower value of $\bar{X} = 20.8 - (1.645)(3) = 20.8 - 4.94 = 15.87$; $\bar{X}_1 = (15.87 - 15.50)/3 = .12$; area beyond $z_{.12} = .4522$.

3. Lower value of $\bar{X} = 20.8 - (2.33)(3) = 20.8 - 6.99 = 13.81$; $\bar{X}_1 = (13.81 - 15.50)/3 = -.56$; area beyond $z_{.12} = 28.77\%$.

4. a) $\sigma_{\bar{X}_1} = \sigma_{\bar{X}_2} = 10/\sqrt{4} = 5$; $\sigma_{\bar{X}_1 - \bar{X}_2} = \sqrt{25 + 25} = 7.071$; critical value: $\bar{X}_1 - \bar{X}_2 = (2.33)(7.071) = 16.475$; $z = 16.475 - (68 - 62)/7.071 = (10.475 - 6)/7.071 = 1.48$; area beyond $z = .0694$; power = 6.94%.

   b) $\sigma_{\bar{X}_1} = \sigma_{\bar{X}_2} = 10/\sqrt{16} = 2.5$; $\sigma_{\bar{X}_1 - \bar{X}_2} = 10/\sqrt{16} = 2.5$; $\sigma_{\bar{X}_1 - \bar{X}_2} = \sqrt{6.25 + 6.25} = 3.536$; critical value: $\bar{X}_1 - \bar{X}_2 = (2.33)(3.536) = 8.238$; $z = 8.238 - (68 - 62)/3.536 = (8.238 - 6)/3.536 = 0.63$; area beyond $z = .2643$; power = 26.43%.

   c) $\sigma_{\bar{X}_1} = \sigma_{\bar{X}_2} = 10/\sqrt{64} = 1.25$; $\sigma_{\bar{X}_1 - \bar{X}_2} = \sqrt{1.563 + 1.563} = 1.768$; critical value: $\bar{X}_1 - \bar{X}_2 = (2.33)(1.768) = 4.119$; $z = 4.119 - (68 - 62)/1.768 = $

$- (1.881) / 1.768 = -1.06$; area beyond $z = .50 + .3554 = .8554$; power $= 85.54\%$.

d) $\sigma_{\bar{X}_1} = \sigma_{\bar{X}_2} = 10/\sqrt{100} = 1.0$; $\sigma_{\bar{X}_1 - \bar{X}_2} = \sqrt{1 + 1} = 1.414$; critical value: $\bar{X}_1 - \bar{X}_2 = (2.33)(1.414) = 3.295$; $z = 3.295 - (68 - 62)/1.414 = (3.295 - 6)/1.414 = -2.735/1.414 = -1.93$; area beyond $z = -1.93 = .50 + .4732 = .9732$; power $= 97.32\%$.

5. a) Since H_0 is false, there is zero probability of falsely rejecting a true null hypothesis.

   b) $\sigma_{\bar{X}_1} = \sigma_{\bar{X}_2} = .6/\sqrt{36} = .1$; $\sigma_{\bar{X}_1 - \bar{X}_2} = \sqrt{.1 + .1} = 0.447$;   critical value: $\bar{X}_1 - \bar{X}_2 = -(1.96)(.141) = -.277$; upper critical value: $(1.96)(.141) = .277$; $z_1 = (-.277 - .02)/.141 = -2.11$; area beyond $z_1 = .0174$; $z_u = .277 - (1.5 - 1.3)/.141 = (.277 - .2)/.141 = .077/.141 = .55$; area beyond $z_u = .2912$

   c) power $= .0174 + .2912 = .3086$; $\beta = 1 - .3086 = .6914$.

6. $N_a = 36/.75 = 48$.

7. $3/5 = .60$.

8. When the correlation is so low that the increase required for significance with reduced degrees of freedom is large relative to the higher t-ratio obtained from a lowered error term.

SELF-QUIZ: TRUE-FALSE

Circle T or F

T   F   1. Statistical tests for independent groups are sometimes more powerful than tests for correlated groups.

T   F   2. Power increases as sample size increases.

T   F   3. The events "H_0 true" and "H_0 false" are mutually exclusive.

T   F   4. The events "H_0 true" and "H_0 false" are independent.

T   F   5. Employing $\alpha = .05$; if the null hypothesis is true, the probability of accepting H_0 equals .95.

T   F   6. Employing $\alpha = .05$; if the null hypothesis is true, the probability of rejecting H_0 equals .05.

T   F   7. Employing $\alpha = .05$; if the null hypothesis is false, the probability of accepting H_0 equals .05.

T   F   8. Employing $\alpha = .05$; if the null hypothesis is false, the probability of rejecting H_0 equals .95.

T   F   9. Power is defined as the probability of rejecting a false null hypothesis.

T   F   10. If the null hypothesis is true, power will increase as we increase the size of the sample.

T   F   11. The power of a test equals the probability of obtaining the critical value of the sample statistic in the distribution under H_0.

T   F   12. The probability of obtaining the critical value of the sample statistic in the distribution under H_0 equals α.

T   F   13. Power is unaffected by changes in α-level.

T   F   14. Power increases as α-level decreases.

T   F   15. β increases as α-level decreases.

T   F   16. The higher the α-level is, the lower is the value of z required to reject H_0.

T   F   17. If H_0 is false, power increases as the critical value of z decreases.

T   F   18. An obtained z which is significant for a two-tailed test may not be significant for a one-tailed test.

T   F   19. A one-tailed test will always be more powerful than its two-tailed alternative.

T   F   20. If the parameter is not in the predicted direction, the two-tailed test is more powerful than the directional H_1.

T   F   21. Nonparametric tests entail less risk of a Type II error than parametric tests.

T   F   22. Parametric tests are always more powerful than nonparametric tests.

T   F   23. As long as the underlying assumptions are met, parametric tests entail less risk of a Type I error than nonparametric tests.

T   F   24. If H_0 is false, the probability of a Type I error is zero.

T   F   25. If H_0 is true, the probability of a Type II error is zero.

T   F   26. A nonparametric test may be made as powerful as a parametric test by employing a larger sample size.

T   F   27. Power efficiency is concerned with the increase in α-level that is necessary to increase the power of a test.

T   F   28. If we employed 10 subjects with Test A, we need to employ 20 subjects with Test B to achieve equal power.

Answers: (1) T; (2) T; (3) T; (4) F; (5) T; (6) T; (7) F; (8) F; (9) T; (10) F; (11) F; (12) T; (13) F; (14) F; (15) T; (16) T; (17) T; (18) F; (19) F; (20) T; (21) F; (22) F; (23) F; (24) T; (25) T; (26) T; (27) F; (28) F.

SELF-TEST: MULTIPLE-CHOICE

1. The probability of a Type I error when $\beta = .05$ is:

   a) .05   b) .95   c) .025   d) .975   e) cannot say without knowing α

2. The power of a test when $\beta = 0.05$ is:

   a) .05   b) .95   c) .025   d) .975   e) cannot say without knowing α

3. For which of the following will the power of the test be greatest?

   a) $N = 5$ b) $N = 10$ c) $N = 25$ d) $N = 250$
   e) insufficient information

4. For which of the following will the power of the test be greatest?

   a) $N = 5, \alpha = .05$ b) $N = 5, \alpha = .01$ c) $N = 50, \alpha = .05$
   d) $N = 50, \alpha = .01$ e) insufficient information

Questions 5–11 refer to the following statistics:
Given two normal populations: $\mu_1 = 50$ $\mu_2 = 54$
$\sigma_1 = 5$ $\sigma_2 = 5$

5. For which of the following will the power of the test be greatest?

   a) $H_1:\mu_1 > \mu_2, \alpha = .01$ b) $H_1:\mu_1 > \mu_2, \alpha = .05$

   c) $H_1:\mu_1 < \mu_2, \alpha = .05$ d) $H_1:\mu_1 < \mu_2, \alpha = .01$
   e) insufficient information

6. A sample of 25 cases is drawn from each population. What is the probability of a Type I error?

   a) 0   b) .01   c) .05   d) 1.00   e) insufficient information

7. A sample of 25 cases is drawn from each population. What is the value of $\sigma_{\bar{x}_1 - \bar{x}_2}$?

   a) 1.00   b) 2.00   c) 1.41   d) .40   e) 0.63

8. A sample of 25 cases is drawn from each population. Employing $\alpha = .01$ (one-tailed test), and

$$z = \frac{(\bar{X}_1 - \bar{X}_2) - (\mu_1 - \mu_2)}{\sigma_{\bar{x}_1 - \bar{x}_2}}$$

   as the test statistic, what is the minimum difference between sample means required to reject $H_0:\mu_1 - \mu_2 = 0$?

   a) 2.33   b) –2.33   c) –7.29   d) 1.41   e) 3.29

9. Employ the information in Multiple-Choice Problem 8 for Problems 9 through 11. What is the power of the test?

   a) .01   b) .19   c) .50   d) .31   e) .69

10. β equals:

   a) .01   b) .10   c) .50   d) .31   e) .69

11. The probability of a Type I error is:

   a) 0.00   b) .0005   c) .01   d) .02   e) .99

12. Test A has a power efficiency of 75% relative to Test B. If for Test B we employed 60 subjects, what is the N required to achieve equal power with Test A?

   a) 60   b) 75   c) 80   d) 15   e) 45

13. Which of the following increases the risk of a Type I error in the comparison of two independent groups?

   a) matching the subjects on a variable that is uncorrelated with the criterion variable and employing the Student t-ratio for correlated samples
   b) employing $\alpha = .01$ instead of $\alpha = .05$
   c) reducing the number of subjects
   d) all of the above
   e) none of the above

14. Which of the following increases the risk of a Type II error in the comparison of two independent groups?

   a) matching the subjects on a variable that is uncorrelated with the criterion variable and employing the Student t-ratio for correlated samples
   b) employing $\alpha = .01$ instead of $\alpha = .05$
   c) reducing the number of subjects
   d) all of the above
   e) none of the above

15. Employing $\alpha = .01$ instead of $\alpha = .05$ will:

   a) increase the risk of a Type I error
   b) increase the risk of a Type II error
   c) increase the power of the test
   d) not affect the power of the test
   e) decrease β

16. Which of the following could be called a "distribution-free test of significance"?

   a) F-test
   b) nonparametric test
   c) HSD test
   d) parametric test
   e) t-test

17. Identify the factor having a direct effect on power:

   a) α-level
   b) sample size
   c) H_1
   d) all of the above
   e) none of the above

18. With regard to the alternative hypothesis, which statement is false?

   a) the critical value of z is lower for a directional test than for a nondirectional one
   b) if the parameter lies in the opposite direction than predicted by H_1, the one-tailed test is less powerful than the two-tailed test
   c) a z that is not significant for a two-tailed test may be significant for a one-tailed test
   d) if H_1 is false, a one-tailed and a two-tailed test are of equal power
   e) none of the above

Answers: (1) e;   (2) b;   (3) d;   (4) c;   (5) c;   (6) a;   (7) c;   (8) c;   (9) e;   (10) d;   (11) a;   (12) c;   (13) e;   (14) d;   (15) b;   (16) b;   (17) d;   (18) d.

17

Statistical Inference with Categorical Variables

BEHAVIORAL OBJECTIVES

Conceptual Objectives

1. What are the necessary conditions under which to use the normal curve for approximations of binomial values?

2. Why has the χ^2 test been described as a "goodness of fit" technique in the single-variable case? What is the null hypothesis for the χ^2 test? State the formula for finding the degrees of freedom in the one-variable case.

3. In the two-variable case of the χ^2 test, specify how the expected frequency for each cell is obtained. What is the formula for the degrees of freedom in this case?

4. Know what the assumption of independence of observations in the χ^2 test is and what happens when this assumption is violated.

Procedural Objectives

1. Conduct a test of significance using the binomial expansion. Use Tables M and N in the appendix to the text.

2. Using z-scores and Table A in the appendix to the text, perform a test of significance.

3. Conduct a test of significance for one- and two-variable cases using the χ^2 test. Know how to consult Table B.

CHAPTER REVIEW

A parametric test of significance is usually a more powerful test than the corresponding nonparametric measure. However, if the assumptions underlying the parametric test are not met, the nonparametric test may well be equally powerful. For instance, if the sample distribution is radically skewed or if the sample size is extremely small, the nonparametric test may be preferable over its parametric counterpart. Another case in which a nonparametric test is more suitable is when the data are nominally scaled.

In this chapter, we shall discuss two nonparametric techniques—the binomial test, which we have mentioned previously, and the χ^2 (chi-square) test. Both the binomial test and the χ^2 test can be used with nominally scaled data, that is, with categorical variables. If we have categorical variables, we are not violating the assumptions of the binomial test or the χ^2 test. A problem in which food products are classified as either vegetable or animal involves a categorical variable. Similarly, a study in which subjects are divided into three body-type groups—endomorphic, mesomorphic, or ectomorphic, according to Sheldon's somatotype theory—uses a categorical variable. In the first example, type of food is the categorical variable; in the second, body-type can be called a categorical variable.

Although we illustrated a binomial sampling distribution in Chapter 11 for $P = Q = 1/2$, we shall now expand the definition of and the possible instances for using the binomial test. In the two-category population, the relationship between Q and P, the proportions of cases in each of the two categories, is defined by $Q = 1 - P$. Assuming that we have only two categories, if .80 of the cases are classified under P, then .20 of them must fall into the Q category. The binomial test can be used with nominal data in which there are only two categories. Although in Chapter 11 we limited our discussion to cases in which $P = Q = 1/2$, the values of P and Q need not be restricted in this manner. Actually, P and Q can assume any values so long as their sum equals 1.00.

County officials in a Montana community wish to estimate the divorce rate from a sample of 20 couples. The officials feel that the divorce rate has not increased over the past five years. At that time, the proportion of marriages ending in divorce (P) was .10 and the proportion of couples remaining married (Q) was .90. If a change has occurred, it is probably in the direction of a higher divorce rate. With this information, let us formally state the problem and use the binomial test as our statistical testing technique.

1. H_0:$P \le .10$

2. H_1:$P > .10$

3. Statistical test: Binomial test, since we are dealing with a two-category population.

4. Significance level: $\alpha = .05$.

5. Sampling distribution: Given by the binomial expansion.

6. Critical region: Consists of all values of x that are so extreme that the probability of their occurrence under H_0 is less than or equal to .05. Note that the critical region is one-tailed.

Once the county officials have formally stated the problem, they collect data from the 20-couple sample and find that 14 of the couples are still married while 6 are not.

The simplest procedure to use in solving this problem is to turn to Table N in the appendix to the textbook. With an N of 20 and a value of P equal to .10, we find that to reject H_0 at the .05 significance level, we must obtain a sample in which 5 or more couples had divorced. Since our obtained value of 6 exceeds the critical value, we reject H_0, and we can assert that the divorce rate has changed over the past 5 years.

Table N may be used in problems in which N does not exceed 49 cases. With an $N \ge 50$, Table N would be of no value. Instead, we would have to determine our probability values by using the normal curve approximation to the binomial. If $P = Q = 1/2$ and $N \le 50$, Table M provides the critical values of x or $N - x$, whichever is larger. For example, with P and Q equal to 1/2 and an N of 16, the critical value at $\alpha = .05$, two-tailed test is 13. You must keep in mind while using Table M that x represents the larger of the two frequencies, in this case 13.

As previously noted, with larger values of N, we may use the normal approximation to the binomial. As N increases and as P and Q become closer in value to 1/2, the binomial distribution approaches the normal distribution. Thus, with larger N's we may frequently use z-scores as our test statistic in the two-category case, rather than the binomial test. If you are in doubt about which statistic is more appropriate with a particular set of data, remember this rule: The product of $(N)(P)(Q)$ should equal at least 9 when either P or Q approaches zero for the normal distribution to be suitable. If P or Q is low and the product of N, P, and Q is less than 9, z-scores should not be used. With a product less than 9 and a two-category population, a better choice of a test statistic would be the binomial test.

When there is a sufficiently large N and NPQ equals or exceeds 9, the normal distribution provides an excellent approximation to the binomial distribution. The test statistic z requires that we determine the number of responses in the P-category, obtain the proportion, NP, expected under the null hypothesis and the standard error of the proportion ($/\sqrt{NPQ}$):

$$z = \frac{x - NP}{\sqrt{NPQ}}$$

in which x = the number of responses in the P category.

Let us suppose you are an entomologist who has been retained by a recently industrialized community to ascertain if increased air pollution has changed the proportion of black versus white months inhabiting your region. Some critics of industrialization have argued that, as in England, the smokier air has provided selective protection for the black moths since they blend better than the white moths with the background and are, consequently, less susceptible to predation. Ten years previously, an in-depth study by a colleague found the proportion of white moths was 60%. Your study reveals that, of 62 moths observed, 26 were white. Was there a significant change over the 10-year period? Use $\alpha = .05$, two-tailed test.

In this example, $x = 26$, $N = 62$, and $P = .60$. Applying the z-score approximation to binomial values, we obtain:

$$\frac{26 - 37.2}{\sqrt{(62)(.60)(.40)}} = \frac{-11.2}{3.857} = -2.90$$

In Table A in the appendix of the text, we find that the area beyond $z = -2.90$ equals .0019. The two-tailed value is, then, $2 \times .0019$ or .0038. This clearly exceeds the .05 significance level. Thus, we may conclude that the proportion of white moths in the community has decreased over the past 10 years.

Now we are ready to take up a statistical test with which we have had no experience as yet. The statistic χ^2 is used with nominally sealed data, or categorical variables, just as the binomial test is. The principle behind the χ^2 test of significance is that the test enables us to decide whether or not a significant difference exists between the observed number of cases falling into each category and the expected number of cases in each category. Of course, our significance testing with χ^2, just as with other statistical tests, is based on the null hypothesis. Since we are testing the differences between observed frequencies and expected frequencies, the test is sometimes described as a "goodness of fit" technique. How well do our observations "fit" our expectations?

Now we shall investigate the specifics of the test itself. The χ^2 formula used to test the null hypothesis is

$$\chi^2 = \sum_{i=1}^{k} \frac{(f_0 - f_e)^2}{f_e}$$

where f_o = the observed number in a given category, f_e = the expected number in that category, and $\sum_{i=1}^{k}$ directs us to sum this ratio over all k categories.

Unlike the binomial test, the number of categories in the χ^2 test can be more than two. In other words, $k \geq 2$. To find the degrees of freedom in the case involving only one variable, we can use the formula df $= k - 1$. If there are 10 categories, the number of degrees of freedom is $10 - 1 = 9$. Table B is the correct table to consult for critical values of χ^2 at various α-levels.

Consider the one-variable problem in which a marketing analysis firm is testing the design of five toys. To do so, the researchers collect the toy preferences of a sample of 30 individuals. Each of the 30 sample members indicates which of the five toys he or she prefers. The following table lists the results of the data-collecting.

Toy	Number of individuals preferring this toy
1	6
2	15
3	3
4	0
5	6

A statement of the null hypothesis in this example would be that all the toys are preferred equally, that is, the expected frequency in each category would be $.2 \times 30 = 6$ respondents. To test the null hypothesis, we should use the formula for χ^2:

$$\chi^2 = \frac{(6-6)^2}{6} + \frac{(15-6)^2}{6} + \frac{(3-6)^2}{6}$$
$$+ \frac{(0-6)^2}{6} + \frac{(6-6)^2}{6}$$

$$\chi^2 = \frac{(0)^2 + (9)^2 + (-3)^2 + (-6)^2 + (0)^2}{6} = \frac{126}{6} = 21$$

With 4 degrees of freedom and an α-level of .05, we find, by consulting Table B for critical values of χ^2 that to reject H_0 we would need a χ^2 value greater than 19.488. Since the obtained χ exceeds this value, we must reject H_0 and assert that the toys are not equally preferred.

We should proceed to more complex applications of χ^2, applications in which there is *more* than one variable. Actually, so far as behavioral scientists are concerned, this is the more frequent application of χ^2. Frequently, researchers are concerned with two or more categorical variables. Our approach to situations involving two or more nominal categories will be much the same as in the one-variable χ^2 condition. However, there is an important distinction: With two or more variables, how do we arrive at the expected frequency values for each category? One characteristic of problems with a plurality of categories is that there is no obvious way of assigning expected frequencies. Although observed frequencies are obtained in the same manner as in the one-sample case, expected frequencies must be determined differently.

So that you will understand the method of arriving at expected frequencies, we shall present a two-variable example. Consider a study in which members of the sample are divided into two groups, introvert or extrovert, on the basis of a personality inventory. An experimental situation is set up to elicit a response of conformity or nonconformity from each of the subjects. In this example, we are dealing with two categorical variables, each variable having two categories or levels.

To show you the results of the experiment, we shall construct what is called a 2×2 contingency table. As you might suspect, the "2×2" description comes from the number of variables versus the number of categories per variable. Following is the contingency table for this problem (hypothetical data):

	Experimental Response		
Personality	Conforming	Nonconforming	Row marginal
Introvert	50	70	120
Extrovert	40	20	60
Column marginals	90	90	180

Let us formally state the problem:

1. H_0: There is no difference between extroverts and introverts in conformity or nonconformity behavior.

2. H_1: There is a difference between extroverts and introverts in conformity or nonconformity behavior.

3. Statistical test: χ^2 test of independence.

4. Significance level: $\alpha = .05$.

5. Sampling distribution: chi-square distribution with one df.

6. Critical region: With one df and an alpha of .05, we find from Table B that our critical region is defined by $\chi^2 \geq 3.84$.

First, we should mention how the degrees of freedom are determined in χ^2 with more than one variable. The formula df $= (r - 1)(c - 1)$ is used, where $r =$ the number of rows and $c =$ the number of columns. Applying this formula, you can easily see that the present problem has $(2 - 1)(2 - 1) = 1$ degree of freedom.

The expected frequencies must be obtained *after* the data have been collected. Unlike the one-variable condition of χ^2, the expected frequencies are determined in the two-variable situation following the data collection. To obtain the expected frequencies for each cell of the contingency table, follow the formula

$$f_e = \frac{\text{row marginal} \times \text{column marginal}}{N}$$

All three values necessary in determining the expected frequency for each cell can be supplied after the gathering of the data. In our example, the expected frequencies for the four cells are $(120)(90)/180 = 60$, $(120)(90)/180 = 60$, $(60)(90)/180 = 30$, and $(60)(90)/180 = 30$. As a check on accuracy, add together the expected frequencies. They should sum to N. Thus, $60 + 60 + 30 + 30 = 180 = N$.

With the addition of another variable over the one-variable case, the χ^2 formula must be expanded in the following manner:

$$\chi^2 = \sum_{i=1}^{r} \sum_{i-1}^{c} \frac{(f_o - f_e)^2}{f_e}$$

Since r and c are used to designate rows and columns, respectively, we must obtain the ratio $(f_o - f_e)^2/f_e$ for each cell in the contingency table. Once we have done this, we must sum the products. Thus, $(50 - 60)^2/60 = 1.67$; $(70 - 60)^2/60 = 1.67$; $(40 - 30)^2/30 = 3.33$; $(20 - 30)^2/30 = 3.33$. In our example, we have four squared ratios to sum. Our four ratios are 1.67, 1.67, 3.33, and 3.33. when we sum these ratios, we obtain the value of χ^2, 10.00.

Recall that our critical region is defined by obtained $\chi^2 \geq 3.84$. Since our value of χ^2 exceeds the critical value, we must reject H_0. In other words, there is a difference between introverts and extroverts in conformity behavior (hypothetical data).

SELECTED EXERCISES

1. For what type of data is the binomial test appropriate?

2. What form does the binomial distribution approach as N increases?

3. How do we determine the expected frequencies in the χ^2 test of independence?

4. If one-fourth of all the students in a particular college are freshmen, what is the probability of selecting at random eight students, none of whom is a freshman?

5. A survey indicates that 60% of the people favor a certain candidate. Employing $\alpha = .05$, what would you conclude, if, in a random sample of four people, all favored that candidate?

6. Suppose that 20% of the students at a particular university are economics majors. What is the probability that, of six students selected at random, at least four will be economics majors? What is the critical value shown in Table N for $\alpha = .05$?

7. The owner of a large supermarket believed that at least 80% of his customers preferred regular milk to skim milk. In a random sample of ten shoppers, four preferred skim milk. Employing $\alpha = .05$, what did he conclude?

8. A student at a large university believes that at least three-fourths of the students smoke. In a random sample of five students, he finds three who smoke, two who do not. What does he conclude?

9. A teacher was interested in determining whether there was a preference among students for very early classes over classes meeting late in the day. In a random sample of 100 students, 60 preferred the early class, 40 the late class. H_0: equal preference for both time periods. Employing $\alpha = .01$, what did he conclude?

10. In a particular county, three-fourths of the people usually vote Democratic, one-fourth vote Republican. This year, an interview of 50 voters yielded 20 who would vote for the Republican candidate, 30 for the Democrat. Employing $\alpha = .05$, what do you conclude?

11. A toothpaste manufacturer selects two random and independent samples of 50 people each. Of the 50 people using Brand A, 20 have no cavities, 30 have at least one. Of the 50 people using Brand B, 25 have no cavities. Employing $\alpha = .05$, what do you conclude?

12. The dean of a large university was interested in determining whether there was a preference among students for particular instructors. The same course was offered at the same time but by three different instructors. In a random sample of 99 students, 24 selected Instructor A, 35 Instructor B, and 40 Instructor C. Employing $\alpha = .01$, what do you conclude?

13. A manufacturer claims that his brand of coffee (Brand A) sells as much as the two leading competitors combined. The manager of a large supermarket decides to take a survey to test this assertion. Of 100 sales, 40 were Brand

A, 30 were Brand X, and 30 were Brand Y. What did the manager conclude concerning the manufacturer's claim?

14. An investigator was interested in determining the relationship, for students, between years at school and feelings about a particular issue. She collected data from 200 students and obtained the following results. Employing $\alpha = .01$, what do you conclude?

	In favor	Against
Freshmen	20	30
Sophomores	25	25
Juniors	30	20
Seniors	35	15

15. An investigator was interested in determining if the ratio of men to women at a particular company was independent of race. Of 100 randomly selected Caucasians, 58 were male, and of 100 blacks, 40 were male. Using $\alpha = .01$, what did she conclude?

16. An investigator was interested in determining if salary grade was independent of sex. Employing the following data, what did he conclude? Use $\alpha = .01$.

Salary grade	Male	Female
1	28	62
2	38	27
3	25	15
4	19	11
5	16	10

Answers:

1. Two-category or dichotomous populations.

2. It takes the form of the normal curve.

3. To determine the expected frequency within a cell, multiply the marginal frequencies common to that cell and divide the product by N.

4. To obtain the answer, calculate the probability that all will be nonfreshmen, i.e., $p = (3/4)^8 = .10$. To obtain $(3/4)^8$ on calculators capable of exponentiation, enter .75 in keyboard and then depress y^x (or x^y or similar variation) followed by 8.

5. Since $(.6)^4 = .1295 > p_{.05}$, we accept H_0.

6. $P_{x>4} = 15p^2q^4 + 6pq^5 + q^6 = .01536 + .001536 + .000065 = .01696$. Critical value in Table N = 4.

7. Let P = hypothesized proportion who prefer skim milk = .20.
   Let Q = hypothesized proportion who prefer regular milk = .80.
   Referring to Table N in the text, we see that x must be equal to or greater than 5 to be significant at the .05 level. Since $x = 4$, we accept H_0.

8. Let $P = 1/4$, $Q = 3/4$. $N = 5$, $X = 2$. Accept H_0.

9. $z = (60 - 50)/\sqrt{(.5)(.5)(100)} = 10/5 = 2.00$; $\chi^2 = (60 - 50)^2/50 + (40 - 50)^2/50 = 2.00 + 2.00 = 4$. Note that, in the two-category case, $z^2 = \chi^2$.

10. $z = 30 - (50)(.75)/\sqrt{(.75)(.25)(50)} = -7.5/\sqrt{9.375} = -2.45$; $\chi^2 = (30 - 37.5)^2/37.5 + (20 - 12.5)^2/12.5 = 1.5 + 4.5 = 6$. Note that, in the two-category case, $z^2 = \chi^2$.

11.

	A	B	Row sum	Expected	frequencies	
Cavities	30	25	55	27.5	27.5	55
No cavities	20	25	45	22.5	22.5	45
Column sums	50	50	100	50	50	100

$\chi^2 = (30 - 27.5)^2/27.5 + \cdots + (25 - 22.5)^2/22.5 = 1.01$. If corrected for continuity (which we don't recommend). $\chi^2 = .65$.

12.

	A	B	C
Number selecting	24	35	40
Number expected	33	33	33

$\chi^2 = (24 - 33)^2/33 + (35 - 33)^2/33 + (40 - 33)^2/33 = 2.455 + 0.121 + 1.485 = 4.061$. Since $\chi^2 \geq 9.21$ is required for significance at $\alpha = .01$ level when df = 2, we fail to reject H_0.

13.

	A	B	C
Number selecting	40	30	30
Number expected	33.33	33.33	33.33

$\chi^2 = (40 - 33.33)^2/33.33 + (30 - 33.33)^2/33.33 + (40 - 33.33)^2/33.33 = 1.335 + 0.333 + 0.333 = 2.001$. Since $\chi^2 \geq 9.21$ is required for significance at $\alpha = .01$ level when df = 2, we fail to reject H_0.

14.

	In favor	Against	Row sum
Freshmen	20 (27.5)	30 (22.5)	50
Sophomores	25 (27.5)	25 (22.5)	50
Juniors	30 (27.5)	20 (22.5)	50
Seniors	35 (27.5)	15 (22.5)	50
Column sums	110	90	200

$\chi^2 = (20 - 27.5)^2/27.5 + \cdots + (15 - 22.5)^2/22.5 = 10.101$. Since $\chi^2 \geq 11.345$ is required for significance at $\alpha = .01$ level when df = 3, we cannot reject H_0.

15.

	Male	Female	Row sum
Whites	58 (49)	42 (51)	100
Blacks	40 (49)	60 (51)	100
Column sums	98	102	200

$\chi^2 = (58 - 49)^2/9 + (42 - 51)^2/51 + (40 - 49)^2/49 + (60 - 51)^2/51 = 1.653 + 1.976 + 1.653 + 1.976 = 7.257$. Since obtained χ^2 exceeds 6.635 at df = 1, we reject H_0. Differential hiring of male and female blacks and whites is practiced.

16.

Salary grade	Male	Female	Row sums
1	28 (45.18)	62 (44.82)	90
2	38 (32.63)	20 (32.37)	65
3	25 (20.08)	15 (19.92)	40
4	19 (15.06)	11 (14.94)	30
5	16 (13.05)	10 (12.95)	26
Column sums	126	125	251

$\chi^2 = (28 - 45.18)^2/45.18 + \cdots + (10 - 12.95)^2/12.95 = 20.721$. Since $\chi^2 \geq$ 13.277 is required for significance at $\alpha = .01$ level when df = 4, we reject H_0 and assert that a greater proportion of males are at the higher ends of the salary grades.

SELF-QUIZ: TRUE-FALSE

Circle T or F.

T   F   1. When large samples are employed, the parametric tests are almost always appropriate.

T   F   2. The binomial distribution is based on a continuous variable.

T   F   3. A negative χ^2 value indicates an effect in the opposite direction.

T   F   4. The binomial test is appropriate only when we are dealing with a two-category population.

T   F   5. Employing $\alpha = .01$, we test $H_0 : P = .25$ for $N = 9$, $x = 5$; we accept H_0.

T   F   6. For any given N, as P or Q approaches zero, the binomial distribution more closely approaches the normal distribution.

T   F   7. In the χ^2 test, the null hypothesis specifies the frequencies observed in each category.

T   F   8. In the χ^2 test, the expected frequencies are never based on the obtained frequencies.

T   F   9. In the 1-degree-of-freedom situation, the expected frequencies must equal at least 5.

T   F   10. When df > 1, the observed frequencies must equal at least 5 in 80% of the cells.

T   F   11. A nonparametric test is always a less powerful test than its parametric counterpart.

T   F   12. The preferred method for an investigator to follow is to select the statistical test most appropriate to his or her data once the data collection is complete.

T   F   13. In a two-category population, the relationship between P and Q can be defined standardly as $P = Q = 1/2$.

T   F   14. Sample sizes must exceed 50 in order for Table M to be appropriate.

T   F   15. When binomial values are being approximated from the normal curve, the probability of a given x equals the probability of its corresponding z-score.

Answers: (1) T; (2) F; (3) F; (4) T; (5) T; (6) F; (7) F; (8) F; (9) T; (10) F; (11) F; (12) F; (13) F; (14) F; (15) T.

SELF-TEST: MULTIPLE-CHOICE

1. In a 2×2 chi-square table the obtained frequency for each cell is 20. Total frequency is 80. The theoretical frequency for the cell in column 1, row 1 is:

   a) 40   b) 80   c) 10   d) 2   e) 20

2. The general rule of thumb for ascertaining the degrees of freedom for all contingency-type tables of r rows, c columns, where the marginal totals are utilized in setting up the expected frequencies (chi-square), is:

   a) $df = (r - 2)(c - 1)$ b) $df = (r - 2)(c - 2)$ c) $df = (r - 1)(c - 1)$
   d) $df = (r)(c) - 2$ e) $df = 2(r - c)$

3. Which of the following is a nonparametric test?

   a) t b) chi-square   c) z d) z for correlated proportions
   e) none of the above

4. In a 2×2 chi-square test, how many degrees of freedom are there?

   a) 0   b) 1   c) 2   d) 3   e) 4

5. In testing H_0 that $P = Q = 1/2$ for a two-category population when $N = 9$, we should employ:

   a) the normal approximation to binomial values
   b) the χ^2 test of independence
   c) the χ^2 one-variable case
   d) the binomial table for $P = Q = 1/2$
   e) none of the above

6. In testing H_0 that P = Q = 1/2 for a two-category population when $N = 60$, we should employ:

   a) the normal approximation to binomial values
   b) Table M
   c) the Student t-ratio
   d) Sandler's A
   e) none of the above

7. To test the null hypothesis of equal preference for three candidates for the same political office, a sample of 510 voters is polled. The test of significance we should employ is:

   a) the binomial expansion
   b) the normal approximation to binomial values
   c) the χ^2 test
   d) the Student t-ratio
   e) none of the above

8. When we employ the data in Multiple-Choice Problem 7, the expected frequency for candidate B under H_0 is:

   a) 255   b) 340   c) 153   d) 170
   d) cannot answer without knowing more about the popularity of candidate B

9. The approximation of the normal curve to the binomial is greatest when:

   a) N is large and $P \neq Q \neq 1/2$
   b) N is small and $P \neq Q \neq 1.2$
   c) N is small and $P = Q = 1/2$
   d) N is large and $P = Q = 1/2$
   e) none of the above

10. Given:

$$p(x) = \frac{N!}{x!(N-x)!} P^x Q^{N-x} \qquad N = 8 \qquad x = 7 \qquad P = Q = 1/2$$

   the probability of x objects in the Q category is:

   a) 1/256   b) 9/256   c) 1/32   d) 9/128   e) none of the above

11. Given the same data as in Multiple-Choice Problem 10, the probability of *at least* seven objects in the Q category is:

   a) 7/256   b) 14/256   c) 9/256   d) 16/256   e) 18/256

12. To employ the normal curve approximation to the binomial, when P is near 0 or 1, the product NPQ should equal:

   a) 5 or greater   b) $\sqrt{5}$ or greater   c) no more than 9
   d) 9 or greater   e) $\sqrt{NP}$

13. In a two-category variable in which $P = Q = 1/2$, the sampling distribution of x (the number of objects in one category) has a mean equal to

   a) $\sqrt{NPQ}$ b) $\sqrt{NP}$ c) $\sqrt{NQ}$ d) NPQ e) NP

14. In a two-category variable in which $P = Q = 1/2$, the sampling distribution of x (the number of objects in one category) has a standard deviation equal to:

   a) $\sqrt{NPQ}$ b) $\sqrt{NP}$ c) $\sqrt{NQ}$ d) NPQ e) NP

15. Given the following table:

	Category		
	A	**B**	**C**
f_0	120	80	100

   and that under H_0, $P(A) = P(B) = P(C)$, the χ^2 value is:

   a) 7.40   b) 8.33   c) 7.71   d) 8.00   e) none of the above

16. In the χ^2 one-variable case, if $N = 1001$ and $k = 4$, the number of degrees of freedom is:

   a) 1000   b) 3000   c) 4   d) 4004   e) 3

17. An appropriate statistical technique for answering the question is, "Is there a difference in the preferences of men and women for three different political candidates?" is:

   a) the χ^2 test of independence
   b) the binomial test
   c) the χ^2 one-variable case
   d) the normal curve approximation to the binomial
   e) none of the above

18. In the χ^2 test of independence, $N = 101$, $r = 3$, $c = 2$, the number of degrees of freedom is:

    a) 100   b) 202   c) 6   d) 3   e) 2

19. Given the following table:

		Category A		
		0	1	
Category B	1	a	b	150
	0	c	d	100
		130	120	250

    the expected frequency in cell a is:

    a) 125   b) .60   c) 78   d) 72   e) 130

20. Which of the following is *not* a necessary assumption in the binomial test?

    a) $P = Q = 1/2$
    b) $P + Q = 1.00$
    c) sampling is random
    d) data may be regarded as representing discrete events
    e) all of the above are necessary assumptions

21. In a specific student population ($N = 2000$), 40% are female. Of the 100 students majoring in psychology, 45% are female. On the basis of this information, which of the following conclusions is correct?

    a) the proportion of females majoring in psychology is significantly greater than the proportion of females in the student population
    b) the proportions of men and of women majoring in psychology are not significantly different from the typical male-female ratio in the total student population
    c) women students are more interested in psychology than men students
    d) men students are more interested in psychology than women students
    e) none of the above conclusions is tenable on the basis of the information provided

22. A drug developed to prevent colds is administered to a group of subjects. In the placebo group 23 contract colds; in the experimental group 18 contract colds. The significance of the difference of these two numbers cannot be tested because:

    a) the values are under 25
    b) the categories are overlapping
    c) the N is inflated
    d) data of this kind are not normally distributed
    e) a basis for computing expected frequencies is lacking

23. On the basis of chance alone, what score is one most likely to obtain on a 100-item, 5-alternative multiple-choice test? (Score is number on right.)

    a) 5   b) 20   c) 40   d) 50   e) 60

24. An instructor asks the 50 students in his class to indicate whether or not they agree with the opinions of three different speakers. The results are as follows:

	Speaker 1	Speaker 2	Speaker 3
Agree	17	29	25
Disagree	33	21	25

In determining whether there is a difference in opinions expressed about the three speakers:

a) $\chi^2 = 6.40$
b) employing $\alpha = .05$, two-tailed test, critical value of $\chi^2 = 5.99$
c) since the obtained $\chi^2 > 5.99$, we reject H_0
d) all of the above
e) none of the above

25. Fifty men and fifty women are asked to indicate their opinions on a particular issue. Assume that the hypothesis being tested is that the expected cell frequencies are equal. The results are as follows, with the expected cell frequencies in parentheses.

	Men	Women
Yes	18 (25)	30 (25)
No	32 (25)	20 (25)

In testing this hypothesis:

a) $\chi^2 = 5.92$
b) employing $\alpha = .01$, critical value of $\chi^2 = 5.41$
c) since the obtained $\chi^2 > 5.41$, we reject H_0
d) all of the above
e) none of the above

Answers: (1) e;   (2) c; (3) b; (4) b; (5) d; (6) a; (7) c; (8) d; (9) d; (10) c; (11) c; (12) d; (13) e; (14) a; (15) d; (16) e; (17) a; (18) e; (19) c; (20) a; (21) b; (22) e; (23) b; (24) e; (25) a.

18

Statistical Inference with Ordinally Scaled Variables

BEHAVIORAL OBJECTIVES

Conceptual Objectives

1. When is the Mann-Whitney U-test the appropriate statistical test? What does the null hypothesis predict about U and U'? Specify the relationship between the obtained values of U and U' and the tabled values found in Tables I_1 through I_4. Identify the various notations used to determine the values of U and U'.

2. What is the effect of failure to correct for ties when using the Mann-Whitney U-test?

3. What conditions are necessary for using the sign test? What does the null hypothesis predict about the direction of the changes in paired scores? State the advantages of the sign test.

4. List the necessary conditions for selecting the Wilcoxon matched-pairs signed-rank test. State the underlying assumptions of the signed-rank test.

Procedural Objectives

1. Conduct a test of significance using the formulas for the Mann-Whitney U-test. Familiarize yourself with the procedure for using Tables I_1 through I_4.

2. Using the sign test and Table M, test for the significance of the null hypothesis.

3. Similarly, conduct a test of significance using the Wilcoxon matched-pairs signed-rank test and Table J.

CHAPTER REVIEW

At times researchers may question whether their data meet the assumptions underlying a parametric test of significance. On these occasions, they might look for a nonparametric statistical test to which the data are better suited. In this chapter, we shall discuss some of these nonparametric tests that exist as alternatives to their parametric counterparts. You should keep in mind, however, that if the assumptions of the parametric test are not violated, it is the more powerful measure.

First on our list is the Mann-Whitney U-test, one of the most powerful of the nonparametric tests. The Mann-Whitney U test is called into play as an alternative to the Student's t-ratio with independent samples. For instance, suppose a behavioral scientist cannot assume homogeneity of population variances. Since this is one of the assumptions underlying the t-ratio, the researcher would be in error to use Student's t-ratio. Instead, he or she should consider the Mann-Whitney test as an alternative measure.

As is indicated by its name, with the Mann-Whitney U-test we are concerned with the sampling distribution of the "U" statistic. Let us discuss exactly what the U-statistic is. Suppose we are dealing with an experimental situation in which we have two sets of scores. If we label one set the "E" scores and the other the "C" scores, we can define U as the sum of the number of times each E score precedes a C score. Similarly, U' is defined as the number of times C scores precede E scores. Note here that U' is the greater sum, that is U' is greater than U. Since U' must be the greater sum, if the sum of the number of times each E score precedes a C score is greater than the sum of the number of times each C score precedes an E score, we would define U as the sum of the number of times each C score precedes an E score. This is because U' must be greater than U.

A fairly simple procedure for determining the values of U and U' in the Mann-Whitney test is, first, to assign ranks to the scores. For E of 3, 8, and 15 and C of 9, 13, and 20, we would rank the scores in the following manner:

Rank	1	2	3	4	5	6
Score	3	8	9	13	15	20
Conditions	E	E	C	C	E	C

After ranking the scores as we have just done, we can use the following formulas to determine U and U':

$$U = N_1 N_2 + \frac{N_1(N_1 + 1)}{2} - R_1$$

and

$$U' = N_1 N_2 + \frac{N_2(N_2 + 1)}{2} - R_2$$

where R_1 = the sum of ranks assigned to the group with a sample size of N_1, and R_2 = the sum of ranks assigned to the group with a sample size of N_2.

Let us consider a hypothetical experiment in which the Mann-Whitney U-test is the appropriate statistical test. Suppose we are comparing self-esteem scores between a group of females and a second group of males. Since the data are ordinal rather than interval, we use the Mann-Whitney test rather than Student's t-ratio. Following are the results of this hypothetical study.

Male		Female	
Self-esteem	Rank	Self-esteem	Rank
15	4	8	1
21	6	10	2
32	8	12	3
40	11	19	5
49	13	25	7
50	14	34	9
52	15	36	10
65	17	48	12
		56	16

Selecting males as group 1, we see that $N_1 = 8$ and $N_2 = 9$. Summing the ranks for the two groups, we find that $R_1 = 88$ and $R_2 = 65$. If we follow the formula for U, we obtain the quantity 20:

$$\begin{aligned} U &= (8)(9) + \frac{(8)(9)}{2} - 88 \\ &= 72 + 36 - 88 \\ &= 20 \\ U' &= (8)(9) + \frac{(9)(10)}{2} - 65 \\ &= 72 + 45 - 65 \\ &= 52 \end{aligned}$$

Consulting Table I_1 ($\alpha = .01$), we see that our obtained U falls within the region of nonrejection as indicated by the border values of 9 and 63. Since our value of U does not fall outside these values, we cannot reject H_0.

The scale of measurement involved when the Wilcoxon matched-pairs signed-rank test is used is not so crude as that of the sign test. Since the Wilcoxon matched-pairs test utilizes more information about the data than does the signed-rank test, it is naturally a more powerful test and should be used whenever possible.

Suppose the activity monitoring in the table below were not so crude a measurement that the sign test would not be our only option. In other words, as well as the direction of the differences between paired scores, we might also make some assessment of the magnitude of these differences. It would then be to our benefit to use the Wilcoxon matched-pairs signed-rank test.

Our procedures on the same set of data would, of course, differ somewhat from our sign-test methods. First, rather than obtaining the signs of the differences, we would calculate the quantitative differences between pairs of scores. Our second step would be to rank these differences according to the absolute value of the difference. Note that we are ranking the differences rather than the scores.

Activity Score

Matched pair	Two days	Two weeks	Difference	Rank of difference
A	30	14	+16	+6
B	12	34	−22	−8
C	19	22	−3	−1
D	38	29	+9	+3
E	10	17	−7	−2
F	14	28	−14	−5
G	25	14	+11	+4
H	20	39	−19	−7
I	11	42	−31	−10
J	8	31	−23	−9

Sum positive = 13

As you have probably noticed, the rank of the difference is assigned according to the absolute value of the difference. However, the sign of the difference is then placed before the rank of the difference. For instance, the rank of 2 corresponds to a difference score of −7, so the negative sign is carried over to make the ranking −2.

On the basis of the T-statistic, we make a decision of whether to accept or reject the null hypothesis. In this case, our null hypothesis leads us to expect that the sum of the negative versus the positive ranks more or less balance to a sum in the neighborhood of zero.

To calculate T, we must determine whether the positive ranks or the negative ranks sum to a smaller absolute value. The smaller value becomes our T-statistic. In this example, the positive ranks equal 13 whereas the negative ranks add up to −42. Since the absolute value of the positive ranks sum is less than the value of the negative ranks, our T-statistic is 13. Turning to Table J in the appendix to the textbook, we see that with $\alpha = .05$, we must have a T-value less than or equal to 8 to achieve significance. Again, we must fail to reject the null hypothesis.

SELECTED EXERCISES

1. In what way does the data appropriate to the Wilcoxon test differ from that appropriate for the sign test?

2. What happens to the value of U when we have ties within a group?

3. In what way does the data appropriate to the Mann-Whitney U-test differ from that appropriate for the Wilcoxon test?

4. What are the assumptions underlying the Wilcoxon test?

5. An investigator was interested in determining whether there was a difference in attitudes on a particular issue between the members of two different church denominations. The results were as follows. What did he conclude? Use $\alpha = .05$, two-tailed test.

Denomination I		Denomination II	
13	37	55	48
67	31	34	47
46	31	52	58
42	23	20	39
22	77	37	
1	50	38	

6. The manager of a large company asked his two foremen to rate ten employees on their efficiency. The ratings were from 1 to 7, where 1 was the poorest rating. Was there a significant difference in the ratings of the two foremen? Use $\alpha = .05$, two-tailed test.

Employee	Foreman A	Foreman B
1	7	6
2	4	4
3	6	6
4	4	4
5	3	5
6	7	6
7	4	6
8	4	2
9	6	4
10	2	3

7. An instructor believes that the students in her class who are majoring in the subject earn better grades than those students who are nonmajors. She randomly selects ten pairs of students of equal ability and compares their final grades. What does she conclude? Use $\alpha = .05$, one-tailed test.

Majors	Nonmajors	Majors	Nonmajors
A	A–	B–	C+
B+	B	A–	C
C	C+	C	D
D	D	C	D
C+	F	B	C

8. An investigator believed that cigarette smokers are more anxious than nonsmokers. He administered an anxiety test to 25 randomly selected subjects. What did he conclude? (The higher the score, the more anxiety.) Use $\alpha = .05$, one-tailed test.

Smokers		Nonsmokers	
23	22	27	20
15	22	14	21
19	24	28	11
30	31	16	13
17		25	10
29		12	16
19		18	20

9. An investigator wanted to know which of two methods resulted in a greater improvement in reading rate. He tested two groups of subjects, matched for initial ability, under the two different methods. What did he conclude? Use $\alpha = .05$, two-tailed test.

Method I	Method II	Method I	Method II
23	19	29	30
25	27	29	29
36	29	36	39
31	23	31	25
36	34	36	31

10. From previous experience, the manager of a large company hypothesizes that Process I resulted in a higher rate of production. He randomly selected 40 employees and obtained the following results. Employing $\alpha = .05$, one-tailed test, what did he conclude?

Process I	Process II	Process I	Process II
34	16	29	18
20	17	37	32
11	15	20	31
39	19	30	30
16	18	31	24
26	13	23	22
27	12	25	22
33	16	14	24
21	13	14	33
30	10	19	28

11. Employing the data in the previous problem, assume that the two groups were matched on a related variable. Compare and explain the difference in the results.

12. A certain diet produces the following results with 15 women. Is the diet effective?

Before	After	Before	After
115	113	132	130
133	130	141	132
126	119	146	140
129	124	144	132
138	139	148	144
140	140	127	119
127	125	130	130
145	147		

Answers:

1. The sign test utilizes only the direction of differences, whereas the Wilcoxon test also considers the magnitudes of differences.

2. Ties across conditions result in a test that is more conservative, that is, less likely to reject H_0 when the null hypothesis is false.

3. The Wilcoxon signed-rank test involves correlated measures, whereas the Mann-Whitney U-test uses independent measures.

4. Both the original measures as well as the differences between measures achieve ordinality.

5. $U = (12)(10) + (10)(11)/2 - 131.5 = 120 + 55 - 131.5 = 43.5$
   $U' = (12)(10) + (12)(13)/2 - 121.5 = 120 + 78 - 121.5 = 76.5$

6. The three tied ranks are dropped. Thus, for the sign test, $n = 7$ and $x = 3$. This is not statistically significant.

7. One pair is dropped because of a tie. Of the remaining 9, 8 are in favor of the hypothesis. At $\alpha = .05$, one-tailed test, the critical value is 8. Since the obtained value equals the critical value, we can reject the null hypothesis majors perform no better.

8. $R_1 = 180$, $n_1 = 11$; $R_2 = 235$, $n_2 = 14$. $U = (11)(14) + (11)(12)/2 - 180 = 154 + 66 - 180 = 40$. The critical value at $\alpha = .05$, one-tailed test, is 46 or less. Since obtained U is less than 46, it lies within the critical region of rejection. Smokers appear to score higher on the anxiety scale.

9. One pair of differences is tied so that final $N = 9$. The negative differences occupy the following ranks: 1, 2.5, and 4. Thus, $T = 7.5$. The critical value at the $\alpha = .05$ level, two-tailed test, is 5. Therefore, we fail to reject H_0.

10. $R_1 = 472.5$, $n_1 = 20$; $R_2 = 457.5$, $n_2 = 20$. $U = (20)(20) + (20)(21)/2 - 472.5 = 400 + 210 - 472.5 = 137.5$. The critical value at $\alpha = .05$, one-tailed test, is 138 or less. Since obtained U is less than 138, it lies within the critical region of rejection. Process I appears to result in a higher rate of production.

11. When treated as matched pairs, the pair tied at 30 is dropped. Thus, the final number of pairs is 19. The ranks of the six negative differences are, respectively: 5, 2, 11.5, 10, 17, 9. The sum of these, T, is 54.5. The critical value at $N = 19$ at $\alpha = .05$, one-tailed test, is 53. Since obtained T exceeds this value, we fail to reject the null hypothesis.

12. The two pairs of ties are dropped so that final $N = 13$. Of the remaining differences, the smallest absolute difference is $|-1|$. A $|-2|$ ties for the next four positions, all of which receive a rank of 3.5. Thus, $T = 1 + 3.5 = 4.5$. The critical value at $\alpha = .05$, two-tailed test, is 17 or less. Since obtained T is less than the critical value, we reject H_0. The diet appears to be effective.

SELF-QUIZ: TRUE-FALSE

Circle T or F.

T   F   1. When the measurements fail to achieve ordinal scaling, we employ the Mann-Whitney U-test.

T   F   2. Given: $N_1 = 10$, $N_2 = 7$, $R_1 = 115$. Thus, $U = 10$.

T   F   3. The sign test assumes that the scale of measurement and the differences in scores achieve ordinality.

T   F   4. The Mann-Whitney U-test is one of the most powerful nonparametric statistical tests.

T   F   5. In the Mann-Whitney U-test, the failure to correct for ties increases the probability of a Type II error.

T   F   6. Given $N = 12$, we obtain $T = 8$. Employing $\alpha = .01$, two-tailed test, we accept H_0.

T   F   7. When we employ the sign test on data that satisfy the assumption of the Wilcoxon test, we increase the risk of a Type II error.

T   F   8. In a study involving 15 pairs of subjects, we obtain the following differences: 10 positive, 3 negative, 2 no difference. Employing $\alpha = .05$, two-tailed test, we accept H_0.

T   F   9. The most common use of the Mann-Whitney U-test is as an alternative to the z-statistic.

T   F   10. If we are dealing with two sets of scores, E and C, in the Mann-Whitney U-test, the definition of U is the sum of the number of times each E precedes a C.

T   F   11. Tables I_1 through I_4 are for use with the T-statistic of the Wilcoxon matched-pairs signed-rank test.

T   F   12. If we intend to use the sign test, we should be able to state that a score of 20 is twice the size of a score of 10.

T   F   13. The Wilcoxon matched-pairs signed-rank test is more powerful than the sign test.

T   F   14. In the Wilcoxon test we should drop any pairs with a zero difference.

T   F   15. The Wilcoxon matched-pairs signed-rank test and the sign test will invariably lead to the same conclusion regarding the null hypothesis.

Answers: (1) F; (2) T; (3) F; (4) T; (5) T; (6) T; (7) T; (8) T; (9) F; (10) T; (11) F; (12) F; (13) T; (14) T; (15) F.

SELF-TEST: MULTIPLE-CHOICE

1. Assume: two conditions, experimental and control; subjects assigned at random to experimental conditions; scale of measurement ordinal or higher; assumption of normality cannot be maintained. The appropriate test statistic is:

   a) Student's t-ratio for uncorrelated samples
   b) Wilcoxon's matched-pairs signed-rank test
   c) the sign test
   d) Mann-Whitney U
   e) Correlated-samples t-ratio

2. Assume: two conditions, experimental and control; matched subjects; scale of measurement is ordinal, in which paired scores indicate only the direction of a difference; assumption of normality cannot be maintained. The appropriate test statistic is:

   a) Student's t-ratio for uncorrelated samples

    b) Wilcoxon's matched-pairs signed-rank test
    c) the sign test
    d) Mann-Whitney U
    e) Correlated-samples t-ratio

3. Assume: two conditions, experimental and control, matched subjects; scale of measurement interval or ratio; assumption of normality is valid. The appropriate test statistic is:

    a) Student's t-ratio for uncorrelated samples
    b) Wilcoxon's matched-pairs signed-rank test
    c) the sign test
    d) Mann-Whitney U
    e) Correlated-samples t-ratio

4. Assume: two conditions, experimental and control, matched subjects; scale is ordinal, in which difference in scores is also ordinal; assumption of normality cannot be maintained. The appropriate test statistic is:

    a) Student's t-ratio for uncorrelated samples
    b) Wilcoxon's matched-pairs signed-rank test
    c) the sign test
    d) Mann-Whitney U
    e) Correlated-samples t-ratio

5. Assume: two conditions, experimental and control; subjects assigned at random, scale of measurement is interval or higher; assumption of normality is valid. The appropriate statistic is:

    a) Student's t-ratio for uncorrelated samples
    b) Wilcoxon's matched-pairs signed-rank test
    c) the sign test
    d) Mann-Whitney U
    e) Correlated-samples t-ratio

6. Given: the scores of two independent groups of subjects in a reaction time study: scale of measurement is ratio; scores are skewed to the right. The appropriate test statistic is:

    a) Student's t-ratio for uncorrelated samples
    b) the Mann-Whitney U
    c) the Wilcoxon's matched-pairs signed-rank test
    d) the sign test
    e) the χ^2 test

7. Given the following scores for two groups: Group E, 8, 12, 15, 17; Group C, 2, 4, 7, 9, 10; and

$$U = N_1 N_2 + \frac{N_1(N_1 + 1)}{2} - R_1$$

U equals:

    a) 7   b) 13   c) −22   d) 2   e) none of the above

8. Given: $N_1 = 5$, $N_2 = 5$, $\alpha = .05$, two-tailed test, and the critical value of $U \leq$ 2 or ≥ 23, we obtain a U of 24. We should:

    a) assert H_1
    b) fail to reject H_0
    c) depends on whether we have obtained U or U'
    d) assert H_0
    e) none of the above

9. Given the following ratings assigned to a group of subjects before and after the introduction of the experimental variable:

Before	15	13	10	9	7	6	5	4	2
After	12	11	6	7	8	3	1	6	0

Employing the sign test, $\alpha = .05$, two-tailed test, we conclude:

   a) reject H_0
   b) accept H_0
   c) the experimental conditions had a small effect
   d) the experimental conditions resulted in lower ratings
   e) none of the above

10. Applying the Wilcoxon matched-pairs signed-rank test to the data in Multiple-Choice Problem 9, we would obtain a T of:

   a) 3   b) 36   c) 4.5   d) 31.5   e) none of the above

11. A *disadvantage* of the sign test when applied to ordinally scaled variables is that:

   a) it is frequently difficult to determine the direction of a change
   b) pairs of measurement must be independent of one another
   c) it increases the risk of a Type I error
   d) it does not utilize any quantitative information inherent in the data
   e) all of the above

12. A *disadvantage* of applying the Wilcoxon matched-pairs signed-rank test to ordinally scaled data is that:

   a) it does *not* utilize information concerning the direction of the differences
   b) it loses sight of magnitudes of differences
   c) pairs of measurements must be independent of one another
   d) all of the above
   e) none of the above

13. The test of significance that employs the binomial sampling distribution to arrive at probability values is:

   a) the sign test
   b) the Wilcoxon matched-pairs signed-rank test
   c) the Mann-Whitney U
   d) Student t-ratio
   e) none of the above

14. An alternative to the Student t-ratio for uncorrelated samples when measurements fail to achieve interval scaling or when one wishes to avoid the assumptions of the parametric counterpart is:

   a) Wilcoxon's matched-pairs signed-rank test
   b) Student t-ratio
   c) the Mann-Whitney U
   d) the sign test
   e) none of the above

15. Given the following:

Rank	1	2	3	4	5	6	7	8	9	10
Condition	C	C	C	E	C	C	E	C	E	E

The Mann-Whitney U and U', respectively, are:

a) 4, 20   b) 9, 15   c) 4, 15   d) 9, 20   e) none of the above

16. Which of the following is a parametric test of significance?

    a) the Mann-Whitney U
    b) Student t-ratio for correlated samples
    c) Wilcoxon's T
    d) the sign test
    e) the binomial test

17. Which statistical test does not belong with the group?

    a) the sign test
    b) Wilcoxon's matched-pairs signed-rank test
    c) t-ratio for correlated samples
    d) Mann-Whitney U
    e) $t = \dfrac{\bar{D}}{s_{\bar{D}}}$

Answers: (1) d; (2) c; (3) e; (4) b; (5) a; (6) b; (7) d; (8) a; (9) b; (10) c; (11) d; (12) e; (13) a; (14) c; (15) a; (16) b; (17) d.

APPENDIX

Decision-Making Chart

I have had many students who quickly mastered the techniques of statistical analysis but foundered somewhat when asked to choose the form of statistical analysis appropriate for a given data set. In truth, the decision concerning the appropriate statistical techniques for analyzing data is as much a part of statistics as the actual computations. The decision about which procedures to use is based on a variety of different considerations—the complexity of the design of the study, the type of numerical data, the number of independent variables and the number of levels or categories of each, the nature of the dependent measure or measures, and so forth. The following chart has been designed to aid you in deciding on the form of data analysis of a given data set. It is also keyed to the various sections of the text *Fundamentals of Behavioral Statistics*, seventh edition. At the outset of the course, this chart will not be of great value. However, if you get into the habit of consulting the chart as you move to higher levels of complexity, its usefulness to you will increase as you progress through the course.

1. What is the purpose of the study?

   a. to assess the effects of an independent variable on a dependent measure (an experiment or quasi-experiment)? Go to 2.
   b. to assess opinions, attitudes, or characteristics of a population based on the results of a sample (a survey)? Go to 20.
   c. to ascertain the relationship between two variables or to predict the status of one variable based on the status of another variable (correlational and/or regression research)? Go to 21.

2. What is the scale of measurement of your dependent variable?

   a. if nominal, go to 3.
   b. if ordinal, go to 9.
   c. if interval/ratio or score data, go to 12.

3. How many independent variables?

   a. if one, go to 4.
   b. if two or more, go to 8.

4. How many categories of the independent variable?

    a. if one, go to 5
    b. if two, go to 6
    c. if more than 2, go to 7

5. This is the one sample case with a nominal scale of measurement. Examples include coin tossing experiments, yes/no responses, and so forth. The obtained results are compared either to a known population value or to a hypothesized population value.

    a. Descriptive statistics: proportions or percentages and standard deviation of a proportion. See Sections 2.7 and 17.3.
    b. Graphing techniques: Bar graph. See Section 3.5.
    c. Inferential statistics: Binomial, Sections 10.2 and 17.2. Normal approximation to binomial values, Section 17.3.

6. This is the single-variable, two-category case with a nominal scale of measurement. Examples include the numbers of voters favoring either of two candidates for political office and proportions of male children born to white versus black parents.

    a. Descriptive statistics: Proportions and percentages. See Section 2.7.
    b. Graphing techniques: Bar graph, Section 3.5.
    c. Inferential statistics: Chi square, one-variable case, Section 17.4.

7. This is the single variable, multicategory case with a nominal scale of measurement. Examples include the numbers of voters favoring one of several candidates for political office and the number of AIDS cases in various risk groups, such as homosexual males, bisexual males, intravenous drug users, recipients of blood transfusions, prostitutes, and heterosexual males and females.

    a. Descriptive statistics: Proportions or percentages. See Section 17.4.
    b. Graphing techniques: Bar graph (Section 3.5) and pie or circle chart, not covered in text but carves a pie or circle into wedges corresponding to the proportions or percentages of measurements in each category.
    c. Inferential statistics: Chi square one-variable case, Section 17.4.

8. This is the two- or multivariable case with a nominal scale of measurement. Examples include hospitalization versus nonhospitalization of children ingesting various classes of household substances containing hydrocarbons.

    a. Descriptive statistics: proportions or percentages, expressed in terms of totals, the column and the row variable(s). See Section 2.7.
    b. Graphing techniques: Bar graph, Section 3.5.
    c. Inferential statistics: Section 10.6 and chi square test of the independence of categorical variables (Sections 17.5 and 17.6).

9. Are the samples independent (different subjects in each group) or correlated (repeat measures on same subjects, subjects formed into pairs on the basis of a variable correlated with the dependent measure, or measurements correlated on the basis of some extraneous variable, such as time of day)?

    a. if independent, go to 10.
    b. if correlated, go to 11.

10. The Mann-Whitney U test may be used with interval/ratio data for which the assumption of normality is questionable or for data that fail to achieve interval/ratio status but can be ranked (ordinal measurement). See Section 18.2.

11. For correlated samples, two possible alternative significance tests are available:

    a. Is it questionable that the values of the variable have any precise quantitative properties other than the algebra of inequalities (greater than or less than)? Use the sign test, Section 18.4.
    b. If we can assume that the scale of measurement is at least ordinal and that the difference in scores achieves ordinality, use Wilcoxon's matched-pairs signed rank test, Section 18.5.

12. Are the samples independent (different subjects in each group) or correlated (repeat measures on same subjects, subjects formed into pairs on the basis of a variable correlated with the dependent measure, or measurements correlated on the basis of some extraneous variable, such as time of day)?

    a. if independent, go to 13.
    b. if correlated, go to 17.

13. How many independent variables are there?

    a. if one, go to 14.
    b. if two, go to 19.

14. How many levels or categories of the independent variable are there?

    a. if one, go to 15.
    b. if two, go to 16.
    c. if more than two, go to 18.

15. You have a single sample design with interval/ratio-scaled data. A comparison is often made with a known or hypothesized population mean.

    a. Descriptive statistics: Measures of central tendency—mean (Section 5.2), median (Section 5.3), and mode (Section 5.4). Also Sections 5.5 and 5.6; Measures of variability—range (Section 6.2), semi-interquartile range (Section 6.3), the mean deviation (Section 6.4), and the variance and standard deviation (Sections 6.5 and 6.6).
    b. Graphing techniques: Histogram for discrete variables (Section 3.7) and frequency curve for continuous variables (Section 3.7 and 3.8).
    c. Inferential statistics: z-statistic when population mean and standard deviation are known (Section 12.3); t-statistic when dealing with unknown parameters (Section 12.5).

16. You have two-sample independent measures design with interval/ratio-scaled data. The usual null hypothesis is that population means from which the samples were drawn are the same.

    a. Descriptive statistics: Measures of central tendency—mean (Section 5.2), median (Section 5.3), and mode (Section 5.4). Also Sections 5.5 and 5.6; measures of variability—range (Section 6.2), semi-interquartile range (Section 6.3), the mean deviation (Section 6.4), and the variance and standard deviation (Sections 6.5 and 6.6).
    b. Graphing techniques: Histogram for discrete variables (Section 3.7) and frequency curve for continuous variables (Sections 3.7 and 3.8).
    c. Inferential statistics: z-statistic when population mean and standard deviation are known (Section 13.1); t-statistic when dealing with unknown parameters (Section 13.3).

17. You have a two-sample correlated measures design or one-way ANOVA with interval/ratio scaled data. The usual null hypothesis is that population means from which the sample were drawn are equal.

a. Descriptive statistics: Measures of central tendency—mean (Section 5.2), median (Section 5.3), and mode (Section 5.4). Also Sections 5.5 and 5.6; measures of variability—range (Section 6.2), semi-interquartile range (Section 6.3), the mean deviation (Section 6.4), and the variance and standard deviation (Sections 6.5 and 6.6).

b. Graphing techniques: Histogram for discrete variables (Section 3.7) and frequency curve for continuous variables (Sections 3.7 and 3.8).

c. Inferential statistics: t-statistic using the correlation between matched pairs (Section 13.6) and the direct-difference method for calculating Student's t-ratio (Section 13.7).

d. ANOVA: One-way randomized block design (Section 14.8).

e. ANOVA: Two-way randomized block factorial design (Sections 15.6–15.7).

18. You have a multigroup independent samples design. There are two or more levels or categories of the independent variable. A one-way analysis of variance is the appropriate inferential technique with the F-ratio being the test statistic (Sections 14.1–14.7).

19. You have a two-way analysis of variance, factorial design. There may be two or more levels or categories of each variable. A two-way analysis of variance is the appropriate inferential technique with the F-ratio being the test statistic. The test includes the effect of the A-variable, the B-variable, and the interaction between the two variables (Sections 15.1–15.5).

20. What is the measurement scale of the dependent variable of the survey?

a. if nominal, go to 3.
b. if ordinal, go to 9.
c. if interval/ratio, go to 12.

21. What is the measurement scale of your two variables?

a. if both are interval/ratio, go to 23.
b. if one or both are ordinal, go to 24.

22. If the relationship appears linear, you should use the Pearson product moment correlation coefficient, r.

a. Descriptive statistics: Pearson r (Sections 8.2 and 8.3); slope of line (b_y or b_x, Section 9.2); the X- and Y-intercept (a_y and a_x, Section 9.2).

b. Inferential statistics: standard error of estimate (Section 9.3), test of significance for Pearson r, one-sample case (H_0:$P = 0$, Section 12.8); testing other hypotheses concerning one-sample case (Section 12.8).

23. Use Spearman r_s, one sample case.

a. Descriptive statistics: Spearman r_s, r_s, or the rank correlation coefficient (Section 8.5).

b. Inferential statistics: Test of significance of r_s, one-sample case (Section 12.8).

MATCHING TEST

After each symbol or equation on the left, place the letter that best matches it from the column on the right.

1. $\dfrac{\sum z_x z_y}{N}$

A. s_{est_y}

2. $1 - \beta$

B. $\hat{s}^2$

3. $s_y \sqrt{1 - r^2}$

C. Standard deviation of a sample

4. $\sum X^2 - \dfrac{(\sum X)^2}{N}$

D. U

5. $100\dfrac{N_a}{N_b}$ percent

E. $X_1 + X_2 + X_3$

6. $\dfrac{SS}{N - 1}$

F. r_x

7. $\sqrt{\dfrac{SS}{N}}$

G. zero

8. $Y' = \bar{Y} + r\dfrac{s_y}{s_x}(X - \bar{X})$

H. $X_{ll} + i\dfrac{(N/2 - \text{cum } f_{ll})}{f_i}$

9. $\dfrac{N!}{x!(N - x)!}P^x Q^{N-x}$

I. SS_x

10. $N_1 N_2 + \dfrac{N(N_1 + 1)}{2} - R_1$

J. Chi square

11. $\sum(X - \bar{X})$

K. r

12. $\sum\limits_{r=1}^{r} \sum\limits_{c=1}^{c} \dfrac{(f_0 - f_e)^2}{f_e}$

L. Sample standard deviation based on unbiased variance estimate

13. $1 - \dfrac{6\sum D^2}{N(N^2 - 1)}$

M. Regression equation

14. $\sqrt{\dfrac{SS}{N - 1}}$

N. Probability of a Type I error

15. $\sum\limits_{i=1}^{3} X_1$

O. z-score

16. r^2

P. $p(x)$

17. Median

Q. Population standard deviation

18. $\dfrac{X - \bar{X}}{s}$

R. Probability of a Type II error

19. σ S. Power of a test

20. α T. Null hypothesis

21. P U. Coefficient of determination

22. $\sigma_{\bar{X}}$ V. $\dfrac{\sigma}{\sqrt{N}}$

23. β W. $1 - Q$

24. $\hat{s}_{\bar{X}}$ X. Power efficiency of Test B

25. H_0 Y. $\dfrac{s}{\sqrt{N-1}}$

Answers: (1) K; (2) S; (3) A; (4) I; (5) X; (6) B; (7) C; (8) M; (9) P; (10) D; (11) G; (12) J; (13) F; (14) L; (15) E; (16) U; (17) H; (18) O; (19) Q; (20) N; (21) W; (22) V; (23) R; (24) Y; (25) T.